Pets

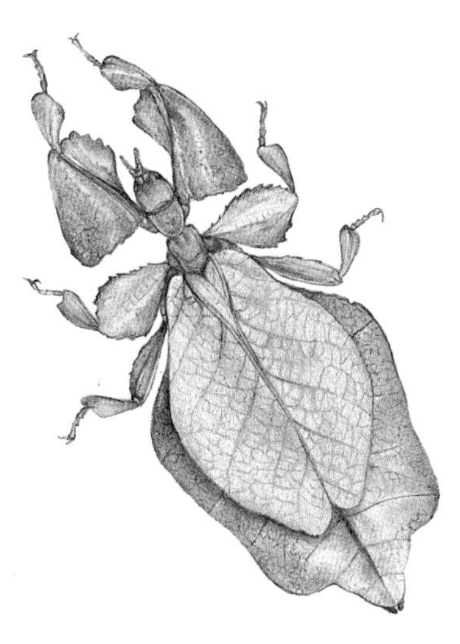

Large Münsterländer

Originating in the Münster region of Germany, the large Münsterländer was developed as a general-purpose hunting dog. Although the breed has been officially recognized in Germany since only the early 1900s, dogs with the appearance of the Münsterländer can be seen in paintings of hunting scenes dating back to the Middle Ages. Characterized by its distinctive thick coat, with feathering on the limbs, this dog displays larger black patches combined with smaller patches of white mottled with black. The dog's handsome head is always black, while its tail ends in a plume. Active and energetic, the Münsterländer has a muscular build and a springy gait. With its love of outings and games, this breed can make an excellent addition to an active family. The Münsterländer is known for being calm, easy to train, and gentle with children.

Scientific name	*Canis familiaris*
Family	Canidae
Size	25–30kg (55–66lb); 58–66cm (23–26in) tall
Distribution	Originated in Germany
Habitat	Domesticated
Diet	High-quality omnivore diet
Breeding	Around 8 puppies twice a year

English Pointer

A pointing breed is a type of gundog bred to find game and "point" it out by aiming its muzzle toward it. The pointer was bred in seventeenth-century England to point small game for greyhounds to chase. As its origins suggest, the pointer is an alert and athletic breed. Coat colors are black, liver, lemon, or orange, in solid colors or speckled on white combined with larger colored patches. Grooming the short coat is not time-consuming. Despite their off-putting loud bark, pointers are usually neither territorial nor aggressive, making them a good choice for a family. As with all sporting breeds, plenty of exercise will be enjoyed, preferably on open land. However, a pointer will be equally happy lounging on the sofa with the family.

Scientific name	*Canis familiaris*
Family	Canidae
Size	16–34kg (35–75lb); 53–71cm (21–28in) tall
Distribution	Originated in England
Habitat	Domesticated
Diet	High-quality omnivore diet: protein, carbohydrate, fats, and nutrients
Breeding	5–6 puppies twice a year

Curly Coated Retriever

The curly coated retriever is one of the less well-known of the retrievers, but it makes a loving and loyal member of the family. Retrievers are gundogs that were bred to retrieve game for the hunter. The curly coated retriever was bred for assisting with hunting waterfowl. Its probable ancestors include the now-extinct English water spaniel and the lesser Newfoundland, making it a lively water dog with good stamina. This retriever's most noticeable feature is its tight, crisp curls, which require considerable grooming attention. Coat colors are black or liver. All curly coated retrievers like to swim and to get plenty of activity and attention. This can also be a calm and laid-back dog when the occasion demands.

Scientific name	*Canis familiaris*
Family	Canidae
Size	23–41kg (50–90lb); 58–69cm (23–27in) tall
Distribution	Originated in England
Habitat	Domesticated
Diet	High-quality omnivore diet: protein, carbohydrate, fats, and nutrients
Breeding	6–8 puppies twice a year

Labrador Retriever

The Labrador is, for good reason, the world's most popular breed of dog. This gentle, good-natured, and intelligent breed makes an excellent family pet. Labradors are known for being patient and affectionate with the whole family, including the very young. The dog's playfulness can occasionally be a little boisterous if it is not well trained, but the Labrador is usually free from the behavioral problems of some other breeds. Although happy in either the city or countryside, the Labrador will always require plenty of exercise. The dog was originally developed to retrieve waterfowl, but today the breed's extensive skills lead to its use as a guide dog, as well as a police and army sniffer dog. The Labrador's hard, weatherproof coat may be black, yellow, or chocolate.

Scientific name	*Canis familiaris*
Family	Canidae
Size	25–36kg (55–80lb); 55–62cm (21.5–24.5in) tall
Distribution	Originated in Canada
Habitat	Domesticated
Diet	High-quality omnivore diet, carefully matched to activity level
Breeding	5–10 puppies twice a year

Golden Retriever

The golden retriever was bred in the nineteenth century to retrieve waterfowl and game birds shot during hunting. As a result, this water-loving breed has a soft, gentle mouth and a dense, waterproof coat. The coat may be straight or fairly wavy and ranges from cream to a rich golden shade. Ancestors of the dog include retrievers, the extinct Tweed water spaniel, the Irish setter, and the bloodhound. The golden retriever is patient, affectionate, and easily trained, perfectly suiting the dog to its frequent usage as a guide dog. This dog becomes very attached to its owners and thrives on plenty of attention and exercise. As with all popular and extensively bred dogs, a golden retriever should always be obtained from a reputable breeder.

Scientific name	*Canis familiaris*
Family	Canidae
Size	28–35kg (62–77lb); 51–61cm (20–24in) tall
Distribution	Originated in Scotland and England
Habitat	Domesticated
Diet	High-quality omnivore diet, carefully matched to activity level
Breeding	6–10 puppies twice a year

English Setter

Setters are gundogs that were bred for use in hunting game birds. The dogs search for game by scent and, on finding their quarry, alert the hunter by freezing into a crouching stance, or set. The English setter is tall and lean, with a handsome and rather noble-looking head. Its silky, speckled coat may be black, liver, orange, or lemon with white and can require considerable grooming to keep clean and presentable. The English setter is known for having a slightly calmer nature than the Irish setter, but can still be mischievous on occasion and requires firm training. Despite this trait, this setter makes a friendly, gentle, and devoted family pet. An energetic dog, the English setter needs plenty of exercise and ideally a large, fenced yard so it can run around.

Scientific name	*Canis familiaris*
Family	Canidae
Size	26–29kg (57–64lb); 61–69cm (24–27in) tall
Distribution	Originated in England
Habitat	Domesticated
Diet	High-quality omnivore diet: protein, carbohydrate, fats, and nutrients
Breeding	6–8 puppies twice a year

Irish Setter

Also known as the Irish red setter, distinguishing this breed from the Irish red and white setter, this is a handsome dog that sports a long, rich chestnut coat. The dog is lean but well muscled, with an aristocratic bearing. Irish setters are highly affectionate and peaceful, with an ability to get along well with children and other pets. The breed is rather unfairly nicknamed the "Mad Irishman," as it has a tendency to a joyful wildness. But good training and plenty of on- and off-lead exercise will allow this dog's zest for life to shine. Highly trained Irish setters are masters of obedience trials, but when allowing a dog a vital off-lead run, it is worth noting that some setters are known to "play deaf" when they are called.

Scientific name	*Canis familiaris*
Family	Canidae
Size	24–32kg (53–70lb); 63–69cm (25–27in) tall
Distribution	Originated in Ireland
Habitat	Domesticated
Diet	High-quality omnivore diet: protein, carbohydrate, fats, and nutrients
Breeding	8–10 puppies twice a year

Clumber Spaniel

It is believed that this spaniel originated in nineteenth-century England, with the breed taking its name from Clumber Park in Nottinghamshire, home of the Duke of Newcastle. The dog's forebears may include older spaniel breeds and the basset hound, as the dog's long back suggests. Like all spaniels, the Clumber was developed for its skill at finding game birds, and has a long coat and low-set ears. The Clumber is a great deal heavier and slower than most spaniels, with its usual walking pace being something of a trundle. Its coat is white with yellow or orange markings. With the loose skin on its forehead and drooping lower eyelids, this breed has a characterful expression. The Clumber makes a gentle and affectionate pet, with a particular talent for finding where food is hidden.

Scientific name	*Canis familiaris*
Family	Canidae
Size	28–38kg (62–84lb); 42–46cm (16.5–18in) tall
Distribution	Originated in England
Habitat	Domesticated
Diet	High-quality omnivore diet, carefully matched to activity level
Breeding	4–6 puppies twice a year

English Springer Spaniel

One of the older breeds of spaniels, the English springer spaniel is probably the ancestor of most modern spaniels. The breed originally served as a hunting dog, used to surprise, or "spring," game birds into the air so that they could be seized by a falcon or hawk. Today, the English springer spaniel makes an affectionate and good-natured family dog. With its high energy levels, this breed needs plenty of brisk walks and would prefer a large yard in which to play endless games of fetch. The English springer spaniel is a relatively tall and fast-paced spaniel with a handsome head and ample pendant ears. The close-lying coat appears in black and white, or liver and white, and is not overly time-consuming to groom.

Scientific name	*Canis familiaris*
Family	Canidae
Size	16–25kg (35–55lb); 43–51cm (17–20in) tall
Distribution	Originated in England
Habitat	Domesticated
Diet	High-quality omnivore diet: protein, carbohydrate, fats, and nutrients
Breeding	4–10 puppies twice a year

Sussex Spaniel

The Sussex spaniel is rarer in its native England than it is in the United States, and is considered an endangered breed by the Kennel Club of Great Britain. The Sussex deserves greater popularity, with its kindly disposition and thoughtful hazel eyes. As its name implies, the breed originated in Sussex, England, in the eighteenth century. Due to its large bones and short legs, the Sussex has a distinctive rolling gait. Its skull is broad, with a furrowed brow and low-set ears. The Sussex's straight or slightly wavy coat appears only in a golden liver. This breed makes an excellent companion dog, particularly for an active family. The dog's temperament is calmer than that of the cocker spaniel, but not so calm as the highly docile Clumber spaniel.

Scientific name	*Canis familiaris*
Family	Canidae
Size	16–23kg (35–50lb); 33–40cm (13–16in) tall
Distribution	Originated in England
Habitat	Domesticated
Diet	High-quality omnivore diet: protein, carbohydrate, fats, and nutrients
Breeding	2–6 puppies twice a year

Cocker Spaniel

There are two distinct breeds of cocker spaniel: the American and the English, both known simply as cocker spaniels in their countries of origin and are similar in appearance. Both descend from English spaniels and are classed as sporting dogs or gundogs. Their name is commonly held to stem from being used to hunt woodcock in England. Both breeds have beautifully expressive dark eyes and pendulous ears. Coats are commonly solid colors (such as black, silver, light cream, brown, dark red, and tan), parti-colored (such as black and white, brown and white, or red and white) or tricolored (such as black and white with tan points, or brown and white with tan points). With its continuously wagging tail, the cocker spaniel is joyful, loving, and trusting in nature—and it adores playing with toys and children.

Scientific name	*Canis familiaris*
Family	Canidae
Size	American: 11–13kg (24–28lb); 36–39cm (14–15.5in) tall
	English: 12–15kg (27–34lb); 38–43cm (15–17in) tall
Distribution	Originated in the United Kingdom or United States
Habitat	Domesticated
Diet	High-quality omnivore diet needed to maintain coat
Breeding	4 puppies twice a year

Italian Spinone

The spinone (pronounced "spee-no-nay") is believed to have originated in the Piedmont region of Italy and may date back to 500 BC. The breed was developed as a general-purpose hunting dog but is today a highly popular family pet, as well as an assistance dog for people with disabilities. The spinone's coat is wiry and close-lying, with a lengthier beard and mustache and a truly distinctive pair of longer, stiffer eyebrows. Coat colors are a variety of shades of brown and orange on a white, brownish-white, or orange-white background. The spinone is docile, affectionate, and known for being patient with children. This is a sensitive and intelligent breed that responds well to training. The spinone likes to have plenty of exercise, and many owners report that it makes a great jogging companion.

Scientific name	*Canis familiaris*
Family	Canidae
Size	29–39kg (64–86lb); 57–70cm (22.5–27.5in) tall
Distribution	Originated in Italy
Habitat	Domesticated
Diet	High-quality omnivore diet: protein, carbohydrate, fats, and nutrients
Breeding	4–13 puppies twice a year

Standard Poodle

The poodle breed may have originated in Germany, as "poodle" comes from the German *pudelhund,* meaning "splashing dog," referring to the dog's water-retrieving roots. France, however, is credited with developing the modern breed and its different sizes. The four sizes are standard, medium, miniature, and toy. Poodles are best known for their thick, curly coat, which is usually seen in a solid color such as black, chocolate, silver, blue, apricot, cream, or white. Unlike most dogs, which have double coats, poodles lack an undercoat. The coat will need regular clipping and grooming. The clips given to family pets can be as fancy as the owner chooses, but are usually less elaborate than those seen on show dogs. With their high intelligence and eagerness to please their owners, poodles are easy to train.

Scientific name	*Canis familiaris*
Family	Canidae
Size	30–34kg (66–75lb); 46–60cm (18–24in) tall
Distribution	Originated in France
Habitat	Domesticated
Diet	High-quality omnivore diet: protein, carbohydrate, fats, and nutrients
Breeding	Around 7 puppies twice a year

Norwegian Elkhound

Although it is classified as a hound in many countries, the national dog of Norway is not related to the hounds and does not hunt like a hound—its role in hunting was to hold moose and other large game at bay until the hunter's arrival. This is a spitz-type dog, characterized by long, thick fur and pointed ears and muzzle, and is usually found in the Arctic regions. An ideal elkhound has a tightly curled tail. This breed is loyal to its pack and makes a good family pet. It loves cold weather and needs daily exercise. An elkhound's intelligence, its strong drive to track, and an independent streak can make it challenging to train. This dog's thick coat and twice-yearly molting can produce drifts of fur. In parts of Norway, this is used to make sweaters.

Scientific name	*Canis familiaris*
Family	Canidae
Size	20–27kg (45–60lb); 46–53cm (18–21in) tall
Distribution	Originated in Norway; temperate to cold climates
Habitat	Domesticated
Diet	High-quality omnivore diet, with vitamin supplements if recommended
Breeding	7–14 puppies twice a year

Bloodhound

Also known as the St. Hubert hound, the bloodhound was bred in about 1000 by the monks of the St. Hubert monastery in Belgium. Its excellent tracking abilities led to the dog being drawn to breed with other scent hounds, such as the English foxhound. The bloodhound is still used to track missing persons today. A gentle and affectionate dog, the bloodhound makes a great family pet—although they need supervising around small children because their bulk could easily knock a child over. Famous for its drooling, the bloodhound can be difficult to train because of its strong interest in tracking scents. The acceptable colors for bloodhounds are black and tan, liver and tan, or red. The average lifespan of a bloodhound is just under seven years, making it one of the shortest-lived purebred dogs.

Scientific name	*Canis familiaris*
Family	Canidae
Size	36–50kg (80–110lb); 58–69cm (23–27in) tall
Distribution	Originated in Belgium
Habitat	Domesticated
Diet	High-quality omnivore diet, two or three small meals a day; plenty of water
Breeding	7–8 puppies twice a year

Basenji

The basenji is most famous for the fact that it does not bark. Due to its unusually shaped larynx, it produces a distinctive yodel-like sound, leading it to be called the "voiceless dog." However, the basenji can mimic sounds and will copy barking if it is raised among barking dogs. A breed of hunting dog and a member of the sight hound family, the basenji originated in central Africa. It is small and elegant, with a tightly curled tail and erect ears. A basenji's forehead is wrinkled and its eyes are almond-shaped, giving the impression of squinting thoughtfully. In behavior and temperament, the basenji has some traits in common with cats: it is very intelligent, but is too independent to respond well to training. It should not be left with noncanine pets and is happiest among older, more considerate children.

Scientific name	*Canis familiaris*
Family	Canidae
Size	9–12kg (21–26lb); 38–43cm (15–17in) tall
Distribution	Originated in Africa; bred mainly in the United States and United Kingdom
Habitat	Domesticated; African rain forests
Diet	High-quality omnivore diet with plenty of green vegetables
Breeding	4 puppies once a year

Pharaoh Hound

The pharaoh hound is the national dog of Malta, where it is called *kelb tal-fenek*, or "rabbit hound." The dog gets its English name from the fact that it was thought to resemble the dogs in paintings on the walls of Egyptian tombs. However, DNA analysis has revealed that the pharaoh is a much younger breed, descended from European hunting dogs. The pharaoh has been classified variously as a sight hound and pariah dog; both groups were bred to chase large game. A graceful and noble-looking dog, the pharaoh has a fine, short coat. The only color accepted by most kennel clubs is red, although the shade can vary. Intelligent and extremely active, the pharaoh needs a good run every day. Its quietly affectionate nature makes this playful dog a good family pet.

Scientific name	*Canis familiaris*
Family	Canidae
Size	18–27kg (40–60lb); 53–64cm (21–25in) tall
Distribution	Malta; rare elsewhere
Habitat	Domesticated
Diet	High-quality omnivore diet: protein, carbohydrate, fats, and nutrients
Breeding	7–8 puppies twice a year

Irish Wolfhound

The tallest dog breed, the Irish wolfhound will be happiest in a large house with access to plenty of space. The breed is very old, possibly bred as a war dog by the Celts as early as the first century BC. The dog's great size, speed, and intelligence once made it ideal for hunting wolves, which is the source of the animal's name. Despite its size, a wolfhound can usually be trusted with children due to its exceptionally sweet and friendly nature. The coat of a wolfhound can be rough or smooth, and is most commonly wheaten or gray, though brindle, red, black, white, brown, and fawn are also accepted by kennel clubs. Sadly, this is not a very long-living dog, with life spans varying between five and nine years.

Scientific name	*Canis familiaris*
Family	Canidae
Size	48–82kg (105–180lb); 74–90cm (29–36in) tall
Distribution	Originated in Ireland
Habitat	Domesticated
Diet	High-quality omnivore diet, two meals a day
Breeding	Around 8 puppies twice a year

Basset Hound

This short-legged but very heavy dog gets its name from the French word *bas,* meaning "low." A French scent hound, the basset was bred to track rabbits and hares. Its curving, erect tail was developed to allow the dog to be visible when hunting in the undergrowth. The dog's most recognizable features, however, are its pendulous ears, which it can occasionally step on. This breed's facial wrinkles and dangling dewlap can give the basset a sad and endearing appearance. The standard colors for the basset are tricolor, in black, tan, and white; red and white; honey and white; and lemon and white. The basset is a placid-natured and friendly dog that will fit in well with the average family. This dog has a hearty appetite, so should be encouraged to take plenty of exercise.

Scientific name	*Canis familiaris*
Family	Canidae
Size	23–29kg (50–65lb); 33–38cm (13–15in) tall
Distribution	Originated in France
Habitat	Domesticated
Diet	High-quality omnivore diet: protein, carbohydrate, fats, and nutrients
Breeding	Around 8 puppies twice a year

English Foxhound

The English foxhound is the rarest of the pet dog breeds, but owners report that their pets are friendly and lively companion animals. The dog was bred for the sport of foxhunting. Foxhounds are traditionally kept in packs and have been selectively bred over centuries to enjoy the chase above all else. As a companion animal, the foxhound is best suited to families with plenty of experience in handling dogs and the time to exercise frequently, preferably in rural areas. The dog needs a great deal of food to sustain its high-energy lifestyle. There is also an American foxhound, descended from an English pack that traveled to America in the seventeenth century. The American breed is taller and slighter than its English counterpart.

Scientific name	*Canis familiaris*
Family	Canidae
Size	25–34kg (55–75lb); 58–68cm (23–27in) tall
Distribution	Originated in England
Habitat	Domesticated
Diet	High-quality omnivore diet: protein, carbohydrate, fats, and nutrients
Breeding	5–7 puppies twice a year

Afghan Hound

The Afghan is best known for its long, silky coat, which requires grooming every day to keep it in a condition worthy of this glamorous and dignified dog. This coat was ideal insulation in the cold mountains of Afghanistan, where the dog was originally bred to hunt hares and gazelles. Afghans come in any color, but many individuals have a black facial mask. The Afghan can be rather aloof and even catlike in its temperament, with a definite bent toward independent-mindedness. This trait can make the breed difficult to train, suiting this dog to an owner with plenty of time and patience, and the ability to show firmness when necessary. With its athletic frame, this breed requires plenty of exercise: the Afghan is happiest when racing across a field with its hair flying.

Scientific name	*Canis familiaris*
Family	Canidae
Size	20–29kg (44–64lb); 63–74cm (25–29in) tall
Distribution	Originated in Afghanistan
Habitat	Domesticated
Diet	High-quality omnivore diet: protein, carbohydrate, fats, and nutrients
Breeding	6–8 puppies twice a year

Greyhound

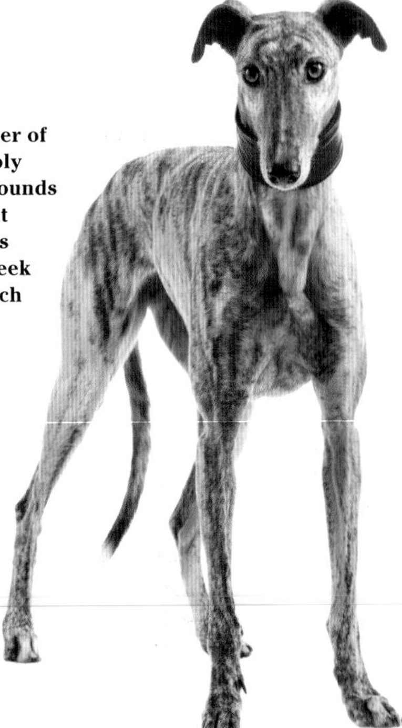

The greyhound is an ancient member of the sight hound group and probably originated in the Middle East. Sight hounds were bred to hunt using keen eyesight and immense speed. The greyhound is the fastest of all dogs: its long legs, sleek build, and deep chest enable it to reach a speed of 72 km/h (45 mph) in just three strides. Greyhounds have very short and easy-to-groom hair, and are found in a wide range of colors and coat patterns. The breed is commonly available from specialty adoption groups, formed to prevent the mistreatment of racing greyhounds and to offer homes to retired racers. Greyhounds make an affectionate and even-tempered pet. Young dogs in particular require plenty of exercise. This is a healthy and relatively long-lived breed, with a life span of up to thirteen years.

Scientific name	*Canis familiaris*
Family	Canidae
Size	27–32kg (60–70lb); 63–74cm (25–29in) tall
Distribution	Originated in the Middle East
Habitat	Domesticated
Diet	High-quality omnivore diet: protein, carbohydrate, fats, and nutrients
Breeding	6–8 puppies twice a year

Otterhound

The otterhound was bred to hunt its semiaquatic quarry on land and in water, so it is equipped with an oily, waterproof coat and webs between its toes. There are only about a thousand otterhounds in the world today, and the breed is considered to be endangered. This is due in part to the outlawing of otter hunting in the United Kingdom in 1978. The Kennel Club of Great Britain has tried to encourage interest in such endangered native breeds to promote their survival. This dog certainly makes a good-tempered and friendly family pet. The ideal family will exercise often and not be fussy about the amount of undergrowth that this shaggy dog will manage to collect in its coat. Garden fences need to be high because these dogs can leap up to 1.5 metres (5 feet) in the air.

Scientific name	*Canis familiaris*
Family	Canidae
Size	40–52kg (88–115lb); 61–71cm (24–28in) tall
Distribution	Originated in England
Habitat	Domesticated
Diet	High-quality omnivore diet: protein, carbohydrate, fats, and nutrients
Breeding	7–10 puppies twice a year

Ibizan Hound

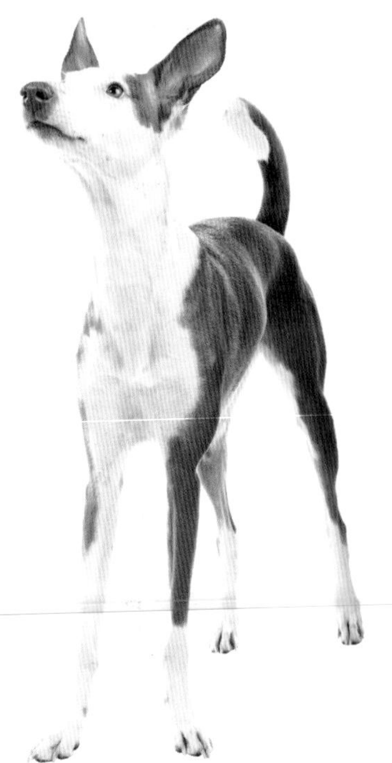

The Ibizan is thought to originate from Muslim Spain, where it was used to hunt small game on the island of Ibiza. There are two types of this fine-boned, elegant hound: the smooth- and the wire-haired. Coat colors are white, chestnut, or tawny, or a combination of these shades. The nose, ears, and eye rims are flesh-colored, while the intelligent eyes are amber. The dog's trademark upright ears taper to a point. The Ibizan has a tendency to skinniness and needs regular feeding to ensure a good layer of fat to offer some protection from cold weather. This dog makes a friendly family pet, with an occasionally entertaining tendency to independent-mindedness. With its strong desire to chase, the Ibizan should be let off the lead only in an enclosed area.

Scientific name	*Canis familiaris*
Family	Canidae
Size	20–30kg (45–65lb); 60–74cm (24–29in) tall
Distribution	Originated in Spain
Habitat	Domesticated
Diet	High-quality omnivore diet: protein, carbohydrate, fats, and nutrients
Breeding	6–12 puppies twice a year

Airedale Terrier

The "king of terriers," so called because it is one of the largest of the terrier breeds, is well known for its humorous and entertaining personality. It loves to be at the center of a family group and expresses its feelings far more readily than more aloof breeds. The Airedale is a stubborn dog, however, and dedication in training is essential, as is socialization with other dogs at a young age. The breed, also known as the waterside terrier, was developed to hunt otters in the river Aire in Yorkshire, England. Like that of many terriers, the Airedale's coat is hard and wiry. The kennel club-recognized colors are a black or dark saddle, with a tan head, ears, and legs. The Airedale's eyes are small, dark, and full of personality.

Scientific name	*Canis familiaris*
Family	Canidae
Size	20–32kg (45–70lb); 56–60cm (22–24in) tall
Distribution	Originated in Yorkshire, England
Habitat	Domesticated
Diet	High-quality omnivore diet: protein, carbohydrate, fats, and nutrients
Breeding	5–12 puppies twice a year

Bull Terrier

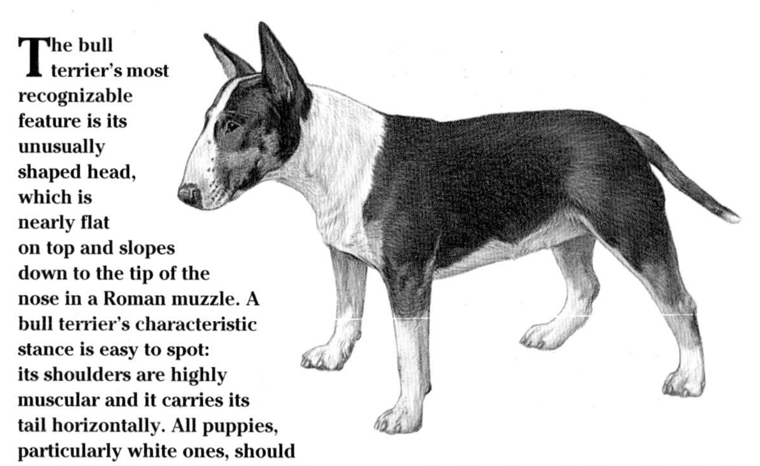

The bull terrier's most recognizable feature is its unusually shaped head, which is nearly flat on top and slopes down to the tip of the nose in a Roman muzzle. A bull terrier's characteristic stance is easy to spot: its shoulders are highly muscular and it carries its tail horizontally. All puppies, particularly white ones, should be checked for deafness (a problem common in white dogs of all breeds). A bull terrier's strength and intelligence need to be exercised frequently. This terrier can be a friendly and playful pet, but can occasionally show aggression to people and animals. As a result of the behavior of poorly trained individuals, the breed has been banned in many countries. The bull terrier was first created by crossing the English white terrier with the English bulldog.

Scientific name	*Canis familiaris*
Family	Canidae
Size	20–33kg (45–72lb); 50–60cm (20–24in) tall
Distribution	Originated in England; banned in many countries
Habitat	Domesticated
Diet	High-quality omnivore diet, maintaining a good balance of exercise and food
Breeding	Around 5 puppies twice a year

Cesky Terrier

This sweet-natured terrier originated in the Czech Republic: its name is pronounced "chess-key" and means simply "Czech terrier." Like all terriers, the Cesky was bred to hunt vermin in their dens. This is a relatively new breed and was recognized only in 1963. The dog is easily identified by its short legs and bushy eyebrows, beard, and mustache. Its triangular ears fold close to the head. Coat colors are various shades of gray-blue and light coffee. Eyes are yellow in brown dogs and brown in gray-blue ones. This terrier is less aggressive with other dogs than some terriers, and its intelligence and quiet personality usually make it easier to train. It loves people, and children are particular favorites. The Cesky is normally patient and loyal, making it an ideal house dog.

Scientific name	*Canis familiaris*
Family	Canidae
Size	6–10kg (13–22lb); 25–30cm (10–12in) tall
Distribution	Originated in the Czech Republic
Habitat	Domesticated
Diet	High-quality omnivore diet: protein, carbohydrate, fats, and nutrients
Breeding	Around 4 puppies twice a year

Lakeland Terrier

The Lakeland terrier originated in the Lake District of northwestern England in the nineteenth century. The breed was developed to hunt down fell foxes, which were a danger to young lambs, and its ancestry includes a variety of older terrier breeds, such as the Dandie Dinmont terrier and Border terrier. Like most terriers, the Lakeland is tenacious, courageous, and alert, and also makes a playful and friendly pet. It likes to get involved in any family activity and enjoys plenty of exercise. This breed has a thick, wiry, and easily groomed coat, which appears in black and tan, blue and tan, liver and tan, or any one of those shades as a single color, plus single-color red or wheaten. The Lakeland has an upright tail and small, dark eyes.

Scientific name	*Canis familiaris*
Family	Canidae
Size	7–8kg (15–18lb); 33–38cm (13–15in) tall
Distribution	Originated in England
Habitat	Domesticated
Diet	High-quality omnivore diet: protein, carbohydrate, fats, and nutrients
Breeding	3–5 puppies twice a year

Norfolk Terrier

The spirited Norfolk terrier is the smallest of the working terriers. With its fun-loving and playful temperament, it will always enjoy participating in family games. Yet the Norfolk is also one of the sweetest-natured of the terriers, thriving on human attention, and is known to be good with children. As the name suggests, the breed was developed in Norfolk, in eastern England, in the late nineteenth century. The Norfolk is related to the equally popular Norwich terrier. The breeds are identical, but while the Norfolk's ears drop forward at the tip, the Norwich sports pricked ears. Both breeds have warm, rough coats in red, wheaten, grizzle gray, or black and tan. The Norfolk has few health issues and a relatively long life expectancy of twelve to sixteen years.

Scientific name	*Canis familiaris*
Family	Canidae
Size	5–6kg (11–14lb); 25–30cm (10–12in) tall
Distribution	Originated in England
Habitat	Domesticated
Diet	High-quality omnivore diet: protein, carbohydrate, fats, and nutrients
Breeding	2–3 puppies twice a year

Scottish Terrier

With its profuse beard and long eyebrows, the Scottish terrier, nicknamed the "Scottie," sports a rakish appearance. The wiry coat usually appears in dark gray or jet black, although wheaten (pale straw-colored) dogs are occasionally seen. Scotties are short-legged and sturdy-bodied with erect ears set on a relatively large head. With its characteristic bright eyes, the Scottie is known for being both friendly and feisty. The breed is sometimes known as the "Diehard" for its courageous and tenacious behavior—though this may sometimes cross into stubbornness. Appropriate training and socialization are needed from a young age to ensure the Scottie does not show aggression to other dogs. Due to its vermin-fighting roots, the Scottie is prone to digging with its large paws and enjoys chasing animals such as squirrels, rodents, and foxes.

Scientific name	*Canis familiaris*
Family	Canidae
Size	8–10kg (17–22lb); 25–28cm (10–11in) tall
Distribution	Originated in Scotland
Habitat	Domesticated
Diet	High-quality omnivore diet: protein, carbohydrate, fats, and nutrients
Breeding	3–5 puppies twice a year

Yorkshire Terrier

With its playful, self-confident personality, this toy terrier deserves its popularity. Often seemingly oblivious to its small stature, the Yorkie, as it is often called, may challenge much larger dogs. Although this breed is a much-loved family pet, its boldness makes it unsuitable for very young children. The Yorkie's jaunty gait and stance are easy to spot: it carries its head and tail high. The kennel club-recognized coat color is blue-black and tan, although many dogs do not conform to this. The dog's long hair, which some owners opt to keep short for easy maintenance, is fine and silky. The breed originated in nineteenth-century Yorkshire, where it was bred to keep vermin under control in mines and textile mills. The Yorkshire terrier was first given its name in 1874.

Scientific name	*Canis familiaris*
Family	Canidae
Size	2–3kg (4.5–7lb); 20–23cm (8–9in) tall
Distribution	Originated in Yorkshire, England
Habitat	Domesticated
Diet	High-quality omnivore diet: protein, carbohydrate, fats, and nutrients
Breeding	Around 4 puppies twice a year

Chihuahua

The tiny Chihuahua originated in Mexico and is named after the northern Mexican state. The dog is believed to be the oldest of the American breeds and is possibly descended from the Techichi, a companion dog favored by the Toltecs. The Chihuahua comes in two types: the smooth-coated (pictured) and the long-coated. Smooth coats are soft and glossy, while long coats are fine and fairly easy to groom. Chihuahuas are spirited and proud dogs, never suffering a perceived insult quietly. Young children should always be supervised with a Chihuahua, for the safety of both parties. Although this dog has a reputation for being highly strung, good training will result in a lovable companion well suited to a city apartment and lifestyle. A Chihuahua is likely to live for twelve to fifteen years.

Scientific name	*Canis familiaris*
Family	Canidae
Size	1–3kg (2–6.5lb); 15–23cm (6–9in) tall
Distribution	Originated in Mexico
Habitat	Domesticated
Diet	High-quality omnivore diet: protein, carbohydrate, fats, and nutrients
Breeding	1–4 puppies twice a year

Italian Greyhound

This miniature greyhound in some ways resembles an exquisite porcelain model: its coat can be given a high shine by a daily rubbing with a piece of silk, but its elegant bones can break fairly easily. Due to its fragility, this dog is not suited to a home with young children. The breed is descended from true greyhounds, with the dog's small stature a result of up to 4,000 years of selective breeding. Despite its name, the breed probably originated in the region of modern-day Greece and Turkey. The Italian greyhound makes a friendly and affectionate companion. It needs plenty of exercise and has a very strong predator instinct, so it should be kept on the lead except in enclosed areas. Its short hair often makes the dog unwilling to take walks in bad weather.

Scientific name	*Canis familiaris*
Family	Canidae
Size	3.5–7kg (8–15lb); 33–38cm (13–15in) tall
Distribution	Originated in Greece and Turkey
Habitat	Domesticated
Diet	High-quality omnivore diet: protein, carbohydrate, fats, and nutrients
Breeding	3–5 puppies twice a year

Maltese

With its long, silky white fur, often tied back in a topknot, the Maltese is one of the most easily recognizable breeds. Grooming the nonshedding and easily dirtied coat can be time-consuming. The dog's expressive, round eyes are surrounded by a darker skin pigmentation known as a "halo." Unlike many toy breeds, the Maltese is not known for excessive barking. This dog is friendly and good-natured, and loves to play with the whole family. As with all dogs, it should be supervised with very young children. Socialization and training at a young age can prevent snappiness. The Maltese is an ancient breed and has been a favored companion throughout history, with owners ranging from Elizabeth I of England to Elizabeth Taylor. The breed is often favored by city dwellers, as it does well in a small home.

Scientific name	*Canis familiaris*
Family	Canidae
Size	1.5–3kg (3–7lb); 20–25cm (8–10in) tall
Distribution	Originated in Malta
Habitat	Domesticated
Diet	High-quality omnivore diet, tailored to activity level
Breeding	2–4 puppies twice a year

Papillon

Papillon means "butterfly" in French, as the white line down this little dog's forehead is said to represent the body of a butterfly, while its tall, plumed ears are the wings. The dog's whole body is covered with long, silky hair that is relatively easy to brush. The papillon is one of the oldest breeds of toy spaniels. Spaniels were originally bred for use in hunting birds, and usually sport drooping ears and a long coat. The papillon was developed as a lapdog and makes an ideal pet for a city dweller. However, these dogs are surprisingly agile and like to get a lot of exercise. This is also an extremely intelligent dog that thrives on plenty of interaction and entertainment. The papillon has a long life expectancy of twelve to sixteen years.

Scientific name	*Canis familiaris*
Family	Canidae
Size	3–4.5kg (7–10lb); 20–28cm (8–11in) tall
Distribution	Originated in Spain and Belgium
Habitat	Domesticated
Diet	High-quality omnivore diet: protein, carbohydrate, fats, and nutrients
Breeding	2–4 puppies twice a year

Pekingese

This toy dog originates from China, where it was once the favored dog for ladies of the imperial court to carry around with them. The Pekingese is renowned for its huge coat, which may be in virtually any color. The exposed skin of the muzzle, lips, and eye rims is always black. Grooming is time-consuming, with a brushing needed every day and a trip to a professional groomer every few months. This is definitely an indoor dog: exercise is not one of its favorite activities. Despite its small size, this dog demands to be respected, and may be aggressive with other animals. But with good training, the Pekingese will make an affectionate and characterful companion. Due to its delicate size, this is not a dog recommended for a home with young children.

Scientific name	*Canis familiaris*
Family	Canidae
Size	4.5–5.5kg (10–12lb); 15–22cm (6–9in) tall
Distribution	Originated in China
Habitat	Domesticated, with temperate climates preferred
Diet	High-quality omnivore diet; avoid overfeeding
Breeding	2–4 puppies twice a year

Shih Tzu

The Shih Tzu is an ancient dog breed that originated in China, probably among the temples of Tibet. The name means "lion dog," referring to the dog's resemblance to the Tibetan snow lion. The breed sports a long, flowing coat that appears in a variety of colors and requires frequent grooming to avoid tangles. A short haircut, known as a "puppy cut," will allow grooming to take place once a month rather than daily. The Shih Tzu has a sturdy body, a short snout, large eyes, and a bannerlike tail that waves high. The short muzzle means that the dog's face may need a wipe after eating, and some dogs need supervision while drinking from a bowl. Renowned for its extroverted, friendly, and proud nature, the Shih Tzu can be excellent company for the whole family.

Scientific name	*Canis familiaris*
Family	Canidae
Size	4.5–7kg (10–16lb); 22–27cm (9–10.5in) tall
Distribution	Originated in China
Habitat	Domesticated, with temperate climates preferred
Diet	High-quality omnivore diet: protein, carbohydrate, fats, and nutrients
Breeding	2–6 puppies twice a year

Cavalier King Charles Spaniel

This pretty and active toy spaniel deserves its great popularity with young families, the elderly, and everyone else in between. The Cavalier holds a long history as a lapdog and has perhaps the most affectionate temperament of all dog breeds. Its love for being around people means that the Cavalier prefers not to be left alone for long periods. Its winning expression and personality mean that firmness will have to be shown to prevent this dog from begging for one too many tidbits for its own good. The Cavalier is a descendant of the slightly more snub-nosed King Charles spaniel, which was the favorite children's pet in the court of the ill-fated English king Charles I. The Cavalier's coat appears in Blenheim (chestnut and white), tricolor (black and white with tan markings), black and tan, and ruby.

Scientific name	*Canis familiaris*
Family	Canidae
Size	4.5–8.5kg (10–19lb); 28–33cm (11–13in) tall
Distribution	Originated in England
Habitat	Domesticated
Diet	High-quality omnivore diet, carefully matched to activity level
Breeding	2–6 puppies twice a year

Shar-Pei

This rare dog is well known for its distinctive wrinkles. In fact, the Shar-Pei comes in two varieties: the Western type is covered in deep wrinkles into adulthood, while the original, Chinese type has skin that appears tighter on its body, although puppies still display deep folds. The dog was first bred in the Guangdong province of China for use as a farm and guard dog. It was originally bred for its strength and scowling face. The breed was later used for fighting, as its unusual coat, with the loose skin and prickly hairs, made it difficult for opponents to get a firm grip with the jaws. Shar-pei means "sand skin" in Cantonese. Although its guard-dog heritage means that the Shar-Pei can be territorial if poorly trained, it makes a devoted family pet.

Scientific name	*Canis familiaris*
Family	Canidae
Size	18–30kg (40–65lb); 45–50cm (18–20in) tall
Distribution	Mostly United States, China, Europe, and Australasia; originated in Guangdong, China
Habitat	Domesticated
Diet	High-quality omnivore diet free from corn, soy, wheat, and gluten to avoid skin problems
Breeding	4–6 puppies twice a year

Bulldog

As its name suggests, the bulldog was bred for use in bullbaiting, a gambling sport popular in England in the seventeenth century. Its short, low body allowed the dog to escape under the bull's horns, while its underbite enabled it to breathe while gripping onto the animal. Due to its muzzle shape, the bulldog is prone to breathing issues and may snore. A bulldog makes a great companion dog, showing extreme loyalty and devotion to its owners. The breed can occasionally show aggression to strangers—human or canine—if provoked. Some bulldogs are so protective that they have been known to intervene when children in the family are being scolded. Compared to other breeds of their size, bulldogs need little exercise and may spend much of the day sleeping.

Scientific name	*Canis familiaris*
Family	Canidae
Size	18–25kg (40–55lb); 28–36cm (11–14in) tall
Distribution	Originated in England
Habitat	Domesticated
Diet	High-quality omnivore diet: protein, carbohydrate, fats, and nutrients
Breeding	Around 4 puppies twice a year

Dalmatian

The distinctive Dalmatian is a large dog with a short white coat covered in black or liver-colored spots. The origins of the breed are disputed, but it is certainly an ancient dog, with its first definite record coming from the Dalmatian coast in Croatia. The Dalmatian was once used as a carriage dog, running between the wheels alongside the horses. In the United States, in the days of horse-drawn fire trucks, the breed was used to run ahead of the horses, clearing the path for the firefighters. The Dalmatian is still a dog that likes to run and needs plenty of vigorous exercise. Its sensitive but exuberant personality means the dog should be supervised around young children, and it requires attentive obedience training. This breed loves people and prefers to stick close to its loved ones.

Scientific name	*Canis familiaris*
Family	Canidae
Size	20–27kg (45–60lb); 48–60cm (19–24in) tall
Distribution	Originated in Croatia
Habitat	Domesticated
Diet	High-quality omnivore diet, low in game and organ meats
Breeding	8–10 puppies twice a year

Lhasa Apso

With its long, hard coat, the Lhasa Apso is ideally suited to the climate of Tibet, where it was bred to act as a monastery guard. Despite its small stature, the dog's keen hearing and intelligence allowed it to perform this task very well. The breed's rich, loud bark gives the impression of a much larger dog. The Lhasa Apso is highly affectionate with its owners but may remain wary of strangers. Its coat, which needs frequent washes and grooming, ranges from golden to gray. Kennel club standards require the Lhasa to have a black nose. It is traditional for the breed to display dark tips on the ears and beard, but many dogs do not show these. The Lhasa is an exceptionally long-lived breed, with some dogs reaching into their early twenties.

Scientific name	*Canis familiaris*
Family	Canidae
Size	6–7kg (13–16 lb); 25–28cm (10–11in) tall
Distribution	Originated in Tibet
Habitat	Domesticated
Diet	High-quality omnivore diet; avoid overfeeding
Breeding	4–6 puppies twice a year

Japanese Spitz

This little dog, with its pure white coat that resembles a snowdrift, was bred in Japan in the late nineteenth century from sled and spitz dogs, including the German spitz, Samoyed, Russian spitz, and American Eskimo dog. The Japanese spitz has the pointed ears, thick coat, and curled-over tail characteristic of spitz dogs, which usually originated in the Arctic regions. The breed is neatly proportioned, with large, oval eyes and a delicately pointed muzzle. The gleaming coat, which is usually odor-free, is surprisingly easy to groom and keep clean. Affectionate and good-natured, this breed thrives on human attention. A well-socialized Japanese spitz is obedient and patient enough to do well around a young family or even other pets. Playing fetch and chasing after frisbees are favorite activities, as are runs in the park.

Scientific name	*Canis familiaris*
Family	Canidae
Size	5–10kg (11–22lb); 30–38cm (12–15in) tall
Distribution	Originated in Japan
Habitat	Domesticated; temperate climate preferred
Diet	High-quality omnivore diet: protein, carbohydrate, fats, and nutrients
Breeding	3–4 puppies twice a year

Boston Terrier

Often referred to as the national dog of the United States, the Boston terrier was developed in Boston in the late nineteenth century and is probably descended from the English bulldog and the English white terrier. The dog was originally bred for fighting but was later developed for companionship, making the modern dog gentle, friendly, and lovable. Most Boston terriers have very strong and determined personalities, but can be trained with patience and firmness. A compactly built body, short tail, erect ears, and short muzzle are the hallmarks of the breed. The short and shiny coat appears in white with black, brindle (gray or tawny tiger stripes), or seal (black with red highlights when seen in bright light). An ideally marked dog displays white on its chest, neck, muzzle, and forehead, and often partway up the legs.

Scientific name	*Canis familiaris*
Family	Canidae
Size	4.5–11kg (10–25lb); 38–43cm (15–17in) tall
Distribution	Originated in the United States
Habitat	Domesticated
Diet	High-quality omnivore diet: protein, carbohydrate, fats, and nutrients
Breeding	3–4 puppies twice a year

Bernese Mountain Dog

The Bernese mountain dog is a member of the working dog group originally used in farming. Its loving and intelligent nature is a strong point of the breed. This dog likes to stick close to its loved ones and will often lean against its owner's leg. The Bernese is usually very sweet with children and often gets along well with other pets. This is a large dog and needs plenty of outdoor activity. The breed is easily recognized by the distinctive tricolor pattern of its coat: black with a white chest and tan markings above the eyes, around the sides of the mouth, the front of the legs, and around the white chest. When viewed from the front in a sitting position, an ideally marked coat will give the impression of a white Swiss cross on the chest.

Scientific name	*Canis familiaris*
Family	Canidae
Size	40–55kg (85–120lb); 60–70cm (24–28in) tall
Distribution	Originated in Switzerland
Habitat	Domesticated
Diet	High-quality omnivore diet: protein, carbohydrate, fats, and nutrients
Breeding	Around 8 puppies twice a year

German Boxer

This stocky, short-haired animal is a working dog and part of the Molosser, or mastiff, group. Its most distinctive features are its square muzzle and underbite. The kennel club-recognized colors for a boxer are fawn and brindle, often with a white underbelly and white on the feet. An energetic breed, the boxer needs plenty of exercise and companionship to keep it from getting bored and amusing itself by chewing or digging. The mature boxer is an excellent jogging companion. This is not usually an aggressive breed, but boxers can be headstrong and need appropriate obedience training and socialization to prevent territorial behavior with other dogs and strangers. Highly trained boxers have been used as guide dogs for the blind and as therapy dogs. The boxer's life span is approximately ten to twelve years.

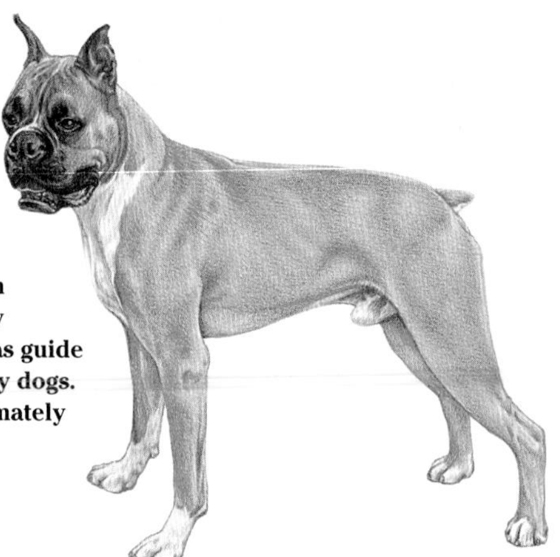

Scientific name	*Canis familiaris*
Family	Canidae
Size	25–32kg (55–70lb); 52–62cm (21–25in) tall
Distribution	Originated in Germany; restrictions on ownership in some countries
Habitat	Domesticated
Diet	High-quality omnivore diet, with two or three small meals daily to prevent bloating
Breeding	6–8 puppies twice a year

Briard

A sheepdog, the rugged briard is also sometimes used in search-and-rescue operations, guarding, and police work. This breed makes a lively and intelligent companion animal that enjoys all the fun of a young family. However, the briard is probably not gentle enough to play with toddlers because it is likely to knock them over. Once this dog has bonded with its family, it is extremely loyal, loving, and protective. As the briard is by instinct a guard dog and herder, it should be socialized with strangers from a young age to ensure it remains outgoing. This dog will even enjoy herding people, if encouraged to do so! The briard needs plenty of exercise and its long, rough, haystacklike coat is demanding to groom. Coat colors are black, gray, and varying shades of fawn.

Scientific name	*Canis familiaris*
Family	Canidae
Size	32–40kg (70–90lb); 55–68cm (22–27in) tall
Distribution	Originated in France
Habitat	Domesticated
Diet	High-quality omnivore diet: protein, carbohydrate, fats, and nutrients
Breeding	8–10 puppies twice a year

Bullmastiff

The bullmastiff was originally bred as a gamekeeper's assistant to track and immobilize poachers. These origins should not be forgotten: this dog is intensely loyal and protective and must be well socialized to encourage friendliness with strangers and other dogs. Good training is essential with this independent-minded and sometimes stubborn breed. This dog should never be left unsupervised with children. However, an experienced and loving owner will appreciate the merits of this powerful but calm dog. The bullmastiff has a hard coat in brindle, fawn, or red. It has a black muzzle and dark markings around the eyes. Cleaning and grooming this neat dog do not take a great deal of effort. The bullmastiff's head shape resembles the nineteenth-century bulldogs from which—along with the Old English mastiff—the dog is descended.

Scientific name	*Canis familiaris*
Family	Canidae
Size	45–60kg (100–130lb); 60–68cm (24–27in) tall
Distribution	Originated in England
Habitat	Domesticated
Diet	High-quality omnivore diet: protein, carbohydrate, fats, and nutrients
Breeding	6–8 puppies twice a year

Doberman Pinscher

The Doberman is a powerful and fast-moving guard dog. Careful breeding has toned down the dog's tendency to a bad temper and it can now make a lively and loyal companion animal. However, this is a dog that needs to know who is in charge, with experienced training and socialization being essential. The breed was developed in the late nineteenth century by a German tax collector, Friedrich Ludwig Dobermann, as a tracking, guard, and police dog. The dog's short coat can take on a high gloss and is usually seen in black with tan on the muzzle, chest, and legs. Other colors occasionally seen are red, blue, or fawn. The dog has a broad chest and a compact, muscular body. A Doberman needs plenty of vigorous exercise and, as a result, it has a hearty appetite.

Scientific name	*Canis familiaris*
Family	Canidae
Size	35–45kg (75–100lb); 60–70cm (24–28in) tall
Distribution	Originated in Germany; restrictions on ownership in some countries
Habitat	Domesticated
Diet	High-quality omnivore diet: protein, carbohydrate, fats, and nutrients
Breeding	Around 8 puppies twice a year

Japanese Akita

First developed to hunt bears and act as a guard dog, the Akita is one of the most ancient breeds. It originated in Japan, probably in the region of the modern Akita prefecture. Two breeds of Akitas are recognized in most countries: the Japanese Akita, also known as the Akita Inu, and the American Akita. The American breed is heavier and comes in a wider range of colors, with a black facial mask. The Japanese comes in only five colors: red, fawn, sesame, brindle, and white. The Akita is a spitz-type dog, characterized by a curled tail, pointed ears, and long, thick fur. The Akita's coat sheds frequently and needs thorough grooming. With its demanding temperament, the Akita is not an ideal family pet. However, with the right owner, this energetic and intelligent dog makes an excellent companion.

Scientific name	*Canis familiaris*
Family	Canidae
Size	34–54kg (75–120lb); 63–71cm (25–28in) tall
Distribution	Originated in Japan
Habitat	Domesticated
Diet	High-quality omnivore diet: protein, carbohydrate, fats, and nutrients
Breeding	3–12 puppies twice a year

Bearded Collie

This highly popular family companion was bred as a herding dog in the Scottish Highlands, but now makes regular appearances at dog shows the world over. The dog's signature shaggy coat requires plenty of grooming to keep it matt-free. The "beardie" comes in gray, blue, black, and sand, with white on the head and lower limbs. Still a working dog at heart, the bearded collie is very energetic and needs plenty of space and activity. A run in the countryside, ideally through as much mud as possible, is the perfect entertainment for this dog. The breed is cheerful and playful, and has beautiful eyes that will make any owner soften after even the most roguish behavior. The bearded collie is fairly long-lived for a breed of its size, with a life span of eleven to fourteen years.

Scientific name	*Canis familiaris*
Family	Canidae
Size	18–27kg (40–60lb); 50–56cm (20–22in) tall
Distribution	Originated in Scotland
Habitat	Domesticated
Diet	High-quality omnivore diet: protein, carbohydrate, fats, and nutrients
Breeding	6–8 puppies twice a year

Border Collie

The Border collie is widely agreed to be the most intelligent dog breed. This herding dog is highly trainable and is a star at obedience competitions. It is still used for handling livestock on farms across the world. The Border collie's most recognizable feature is its keen and alert expression, rather than any physical characteristic. Its coat comes in a range of colors with white, although black with white is the most common. The Border collie needs to be kept both physically and mentally active, so will do best in a grown family with plenty of activity and dedicated attention. By instinct a herder and motivator, this collie may even try to control the family's movements, unless its energies are focused by appropriate training and stimulation.

Scientific name	*Canis familiaris*
Family	Canidae
Size	18–25kg (40–55lb); 46–56cm (18–22in) tall
Distribution	Originated on the borders of England and Scotland
Habitat	Domesticated
Diet	High-quality omnivore diet: protein, carbohydrate, fats, and nutrients
Breeding	4–8 puppies twice a year

Smooth Collie

The smooth collie is the sister dog of the rough collie: the breeds differ only in their coat length. The smooth collie has a short and flat coat, while the rough collie—made popular by the *Lassie* films and television series—has a long and coarse outercoat. Common coat colors are sable, tricolor (black with tan and white), and blue merle (silvery gray marbled with black and tan), all marked with white on the chest, neck, legs, and tail tip. Both the smooth and rough collie make friendly family pets. They are highly intelligent, while their eagerness to please makes them easy to train. An intense love for humans gives these dogs near-telepathic abilities, according to some owners—although they may not be as skilled at finding children down wells as every Lassie lover would wish.

Scientific name	*Canis familiaris*
Family	Canidae
Size	20–34kg (45–75lb); 56–66cm (22–26in) tall
Distribution	Originated in Scotland
Habitat	Domesticated; temperate climate preferred
Diet	High-quality omnivore diet: protein, carbohydrate, fats, and nutrients
Breeding	4–6 puppies twice a year

Komondor

With its long, white, corded coat—which reaches the ground in some adults—the giant komondor is hard to miss. It is also hard to keep, unless an owner is an experienced dog handler and has plenty of outdoor space for this energetic dog to race in. The dog was bred as a livestock guardian in Hungary, with the thick coat protecting the animal from a wolf's bite. An explosion in the coyote population has led to a resurgence in popularity of the komondor as a flock guardian in the United States. Like all guardian dogs, the komondor is calm and peaceful unless it suspects there is danger to its family or herd, making it wary of strangers. Controlling a dog of this size, when roused, is not for the faint of heart.

Scientific name	*Canis familiaris*
Family	Canidae
Size	36–50kg (80–112lb); 60–81cm (23.5–32in) tall
Distribution	Originated in Hungary
Habitat	Domesticated
Diet	High-quality omnivore diet: protein, carbohydrate, fats, and nutrients
Breeding	3–10 puppies twice a year

Hungarian Puli

The puli is a Hungarian sheepdog whose ancestry may date back as far as 6,000 years. The breed's long, enveloping coat is in tight curls that form into cords, making it extremely warm and virtually waterproof. These trademark cords take a great deal of care to keep clean and neat. Coats are a solid color, in black, white, gray, or cream. Some cream pulis have a black facial mask. As a working dog, a puli is dedicated, energetic, and highly intelligent. As a pet, a puli is friendly and playful. It tends to see itself as the guardian of the family, and may view strangers as a threat and move toward overprotective behavior unless well trained. A puli should be obedience trained when young, and requires an owner who knows how to show it who's the boss.

Scientific name	*Canis familiaris*
Family	Canidae
Size	10–15kg (23–33lb); 38–46cm (15–18in) tall
Distribution	Originated in Hungary
Habitat	Domesticated
Diet	High-quality omnivore diet: protein, carbohydrate, fats, and nutrients
Breeding	4–7 puppies twice a year

Hungarian Kuvasz

Although regarded as an ancient Hungarian breed, the kuvasz's true origins may lie as far away as Tibet. What is known for sure is that, around 2000 BC, the Magyar tribes from the Ural region began their slow migration across the central Asian steppes to Hungary, bringing with them kuvasz-type dogs, which served as livestock guardians. The root of the dog's name may be the Turkic word *kuwasz*, meaning "protector." The kuvasz is of a strong build and sports a striking white coat. As with all intelligent guardian dogs, training and socialization are essential. While being loyal, patient, and even humorous, the kuvasz can also be disconcertingly independent and fiercely protective. Many dogs of this breed are prone to barking at real or imagined threats, making them less than popular with neighbors.

Scientific name	*Canis familiaris*
Family	Canidae
Size	35–52kg (77–115lb); 66–76cm (26–30in) tall
Distribution	Originated in Hungary
Habitat	Domesticated
Diet	High-quality omnivore diet: protein, carbohydrate, fats, and nutrients
Breeding	7–8 puppies twice a year

Leonberger

This good-natured giant is known for being easygoing and friendly. Despite its undemanding—and even rather ambling—approach to exercise, the Leonberger is of a size that suits it best to a firm owner with a large house and garden. The Leonberger has a highly water-resistant coat, which sheds heavily. Males often have a mane of very thick fur on the neck and chest. Coat colors are sandy, golden, red, and reddish-brown, with a dark mask on the face. The dog was first bred in the early nineteenth century in the town of Leonberg in southwestern Germany, from the Newfoundland, Saint Bernard, and Pyrenean mountain dog. The dog's great strength was used to pull loads and guard livestock. Only eight members of the breed survived World War II, and these are the ancestors of all of today's Leonbergers.

Scientific name	*Canis familiaris*
Family	Canidae
Size	40–75kg (90–165lb); 64–81cm (25–32in) tall
Distribution	Originated in Germany
Habitat	Domesticated
Diet	High-quality omnivore diet: protein, carbohydrate, fats, and nutrients
Breeding	6–14 puppies twice a year

Alaskan Malamute

This sled dog is powerfully built and sports a short, dense coat suited to subzero temperatures. The malamute exerts its pulling ability even on the end of a lead, so it requires a strong and in-control owner. Obedience training and socialization are essential at an early age to encourage this challenging but highly engaging breed to fit into the average household. The breed is normally friendly with people but can behave territorially with other dogs. The malamute's coat colors range from pure white to white with gray, sable, black, or red. This dog likes to enjoy the snow in winter and to cool down in a sprinkler or pond in the summer. In addition to having a rather wolflike appearance, the malamute has a howl that is difficult to distinguish from that of the wolf.

Scientific name	*Canis familiaris*
Family	Canidae
Size	34–38kg (75–85lb); 58–64cm (23–25in) tall
Distribution	Originated in Alaska and Arctic Canada
Habitat	Domesticated; not suited to hot climates
Diet	High-quality omnivore diet: protein, carbohydrate, fats, and nutrients
Breeding	4–10 puppies twice a year

Mastiff

Often known as the Old English mastiff, this giant dog requires a determined and knowledgeable owner. The breed is powerfully built, with a massive body, broad skull, and large jaws. It is one of the heaviest dog breeds. The mastiff's short coat is apricot-fawn, silver-fawn, fawn, or dark fawn-brindle, with black on the ears, muzzle, and around the eyes. It is believed that the breed is Britain's oldest, and it was once used in baiting games and dogfighting. Fortunately, the mastiff has a calm and good-natured temperament. Like all guard dogs, the mastiff is intensely loyal to its owners and can be territorial with strangers. Although it is known for being good with children of its own family, a mastiff should never be left unsupervised with young children.

Scientific name	*Canis familiaris*
Family	Canidae
Size	80–86kg (175–190lb); 71–76cm (28–30in) tall
Distribution	Originated in England
Habitat	Domesticated
Diet	High-quality omnivore diet: protein, carbohydrate, fats, and nutrients
Breeding	2–5 puppies twice a year

Neapolitan Mastiff

This devoted guard dog is probably a descendant of war and fighting dogs used by the ancient Greeks and Romans. The Neapolitan mastiff is characterized by loose skin over its entire body, with particularly deep wrinkles and folds on the head, and a large dewlap. The dog also has remarkably pendulous lips. As with most loose-jowled dogs, this mastiff is prone to dribbling. With an emphatic shake of the head, it will shower everyone in the vicinity with drool. Like other mastiffs, the Neapolitan is massively built, with a square-shaped head and muzzle. The coat is black, blue-gray, or brown. The loving but dominant personality of the Neapolitan is probably not suited to the average family. This is a dog that must be socialized and trained at a young age to prevent overprotectiveness.

Scientific name	*Canis familiaris*
Family	Canidae
Size	50–70kg (110–155lb); 66–76cm (26–30in) tall
Distribution	Originated in Italy
Habitat	Domesticated
Diet	High-quality omnivore diet: protein, carbohydrate, fats, and nutrients
Breeding	6–12 puppies twice a year

Australian Shepherd

Despite its name, this working dog was developed on ranches in the western United States in the nineteenth century. The name is believed to have been acquired from the dog's association with migrant Australian shepherds. Coat colors for the "Aussie" are black, red, blue merle (marbled black and gray), and red merle (marbled red and silver or buff), with copper points or white markings on the face, chest, and legs. Most Aussies are energetic dogs that love activity, whether it is working, competing in obedience and agility competitions, or being kept thoroughly occupied by their owner. For a more family-oriented dog, look for a breed line that has been developed for a quieter and affectionate personality. Such dogs are renowned for their close bonds with their loved ones and their desire to stick close to them.

Scientific name	*Canis familiaris*
Family	Canidae
Size	16–32kg (35–70lb); 46–58cm (18–23in) tall
Distribution	Originated in the United States
Habitat	Domesticated
Diet	High-quality omnivore diet: protein, carbohydrate, fats, and nutrients
Breeding	6–9 puppies twice a year

Belgian Shepherd

The Belgian shepherd, also known as the Belgian sheepdog, is a tall and elegant dog with the ability to move extremely fast when needed. There are three coat types, with different color patterns. The two types with long, straight topcoats are the Tervuren, which is gray, fawn, or red with a black overlay (pictured), and the Groenendael, which is black. The Malinois has a smooth coat and has the same colors as the Tervuren. The distinctive Laekenois has a wiry coat and is reddish fawn. In the United States, the types are recognized as separate breeds, with only the Groenendael being known as the Belgian sheepdog. All these dogs are highly intelligent and sensitive animals, forming very strong bonds with humans. Like all herding breeds, they need activity and stimulation, as well as appropriate training.

Scientific name	*Canis familiaris*
Family	Canidae
Size	30–34kg (65–75lb); 61–66cm (24–26in) tall
Distribution	Originated in Belgium
Habitat	Domesticated
Diet	High-quality omnivore diet: protein, carbohydrate, fats, and nutrients
Breeding	6–10 puppies twice a year

Newfoundland

Owning a Newfoundland, or "Newfie," is rather like owning a giant and affectionate teddy bear. This breed is renowned for its docile and family-oriented personality. The dog developed on the island of Newfoundland, now in Canada, and is probably descended from an indigenous breed. This is a water dog, with an oily, waterproof coat and webs between its toes, making it a very strong swimmer. The breed was traditionally used by fishermen to pull nets and equipment. Newfies have a famous propensity for carrying out dramatic and life-saving water rescues. Owners report that this breed may also attempt to "rescue" any unsuspecting human swimmers. Newfie colors are black, brown, or white with black markings. This large dog is best suited to a spacious home, preferably with access to water.

Scientific name	*Canis familiaris*
Family	Canidae
Size	50–70kg (110–155lb); 64–71cm (25–28in) tall
Distribution	Originated in Canada
Habitat	Domesticated
Diet	High-quality omnivore diet: protein, carbohydrate, fats, and nutrients
Breeding	8–10 puppies twice a year

German Pinscher

This ancestor of the Doberman pinscher and the miniature pinscher is midway between the two in size. The German pinscher is itself descended from early herding and guard dogs, giving it an alert and intelligent nature, combined with a sharp bark. Pinschers can be distrustful of strangers but tend to make good family members, provided they have been well-bred and trained. They like to be close to the people they love and to be given plenty of exercise and attention. The breed sports a short, dense coat that can take on a good shine with minimal grooming. Colors are tan with black, red, blue, or fawn. A pinscher's body is neat and well-muscled while its features are delicate and expressive. This dog's characteristic gait is light and athletic.

Scientific name	*Canis familiaris*
Family	Canidae
Size	11–20kg (25–45lb); 43–51cm (17–20in) tall
Distribution	Originated in Germany
Habitat	Domesticated
Diet	High-quality omnivore diet: protein, carbohydrate, fats, and nutrients
Breeding	5–8 puppies twice a year

Pembroke Welsh Corgi

Famously kept by Queen Elizabeth II, who had as many as sixteen at one time, the Pembroke Welsh corgi is a popular family dog. The smallest of the herding dogs, the breed was developed in Wales for managing sheep, horses, and cattle by nipping at their heels. Its short stature allows kicks to pass over the dog's head. There are two breeds of Welsh corgi: the Pembroke and the lesser-known Cardigan. Both breeds are sturdily built with long bodies and pricked ears. Pembrokes are usually seen in red and white coats, although sable, fawn, or black and tan dogs are also seen. The Cardigan has a longer tail and appears in a wide variety of colors. Both breeds make active, intelligent, and loyal pets, but have a slight tendency to overeating.

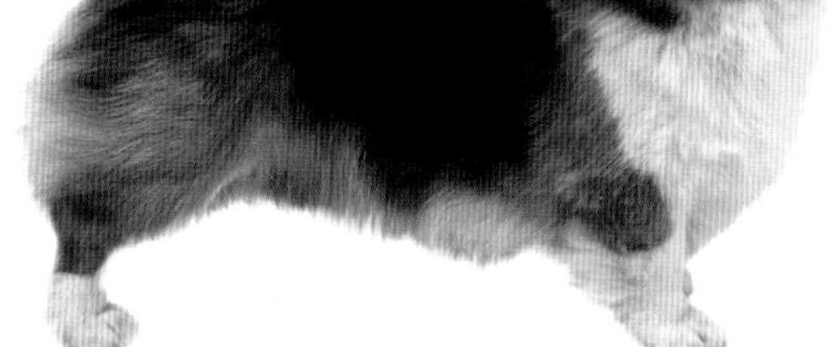

Scientific name	*Canis familiaris*
Family	Canidae
Size	10–12kg (22–26lb); 25–30cm (10–12in) tall
Distribution	Originated in Wales
Habitat	Domesticated
Diet	High-quality omnivore diet, carefully matched to activity levels
Breeding	5–8 puppies twice a year

British Shorthair

Also known as the European shorthair, this is the most popular breed in the United Kingdom. Originally developed from ordinary domestic cats, the breed was later infused with Persian blood to improve the thickness of its coat. British shorthairs exemplify the "cobby" build, with a short, sturdy body shape. The face is round and chubby-cheeked with large eyes, giving the breed what is sometimes described as a teddy bear–like appearance. The short, plush coat is seen in many colors and patterns. The most popular color is blue, known as British blue, but white, black, red, cream, chocolate, lilac, cinnamon, and fawn are also accepted. Patterns include tabby, tortoiseshell, pointed, spotted, and bicolor. British shorthairs are easygoing and undemanding, potentially suiting them to apartment living or to owners who are out during the day.

Scientific name	*Felis catus*
Family	Felidae
Size	3.6–8kg (8–18lb)
Distribution	Originated in the United Kingdom
Habitat	Domesticated
Diet	Meat, fish, and some vegetables
Breeding	Around 5 kittens up to three times a year

Exotic Shorthair

Around 1960, U.S. breeders crossed the American shorthair with the Persian: the exotic shorthair, or "Lazy Man's Persian," was born. It may appeal to people who like the gentle personality of the Persian but prefer a short-haired cat. The exotic shorthair's snub nose; large, round eyes; and small ears give it an extremely cute appearance. The fluffy coat is slightly longer than that of other short-haired breeds and may be pointed (with a pale body and darker extremities), golden, tortoiseshell, blue, or tabby. An ideal choice for city dwellers, the exotic shorthair is a devoted lap cat. It may need reminders of its owners—such as a radio playing—when it is left alone. Unlike the Persian, the exotic shorthair is able to keep its fur tidy with little assistance, although a weekly brushing is advisable.

Scientific name	*Felis catus*
Family	Felidae
Size	3.4–6kg (7.5–13lb)
Distribution	Originated in the United States
Habitat	Domesticated
Diet	Meat, fish, and some vegetables
Breeding	Around 4 kittens up to three times a year; kittens may have long or short hair

Manx

Known for being tailless, the Manx originated on the Isle of Man, off England's west coast. The lack of tail is caused by a naturally occurring mutation of the spine. This dominant gene became prevalent among cats on the island due to the limited breeding population. In fact, Manx cats can range from tailless, known as the rumpy type, through risers and stumpies to fully tailed, known as longies. Only rumpies and risers are seen in shows. The vertebral column of the Manx also differs from that of other cats, giving it a curved back when seen in profile. A wide range of coat colors and patterns is accepted in the Manx. These cats are powerful hunters and do not give the impression of suffering from their lack of a tail.

Scientific name	*Felis catus*
Family	Felidae
Size	3.6–6kg (8–13lb)
Distribution	Originated in the United Kingdom
Habitat	Domesticated
Diet	Meat, fish, and some vegetables
Breeding	Around 3 kittens up to three times a year

American Shorthair

The American shorthair is descended from cats taken to the continent by the first European settlers. In its early years, the breed was allowed to develop without much human interference and largely left to fend for itself, making it robust, versatile, and densely coated. Early in the twentieth century, a selective breeding program was established to develop the cat's best characteristics. The main differences between the American shorthair and its European counterparts is that the American has a less-rounded face and is marginally larger. The breed is recognized in more than eighty colors and patterns, with perhaps the most popular being the silver tabby, displaying black markings on a silver base. American shorthairs are known for making intelligent and active pets, and enjoy exploring outdoors.

Scientific name	*Felis catus*
Family	Felidae
Size	3.6–6.8kg (8–15lb)
Distribution	Originated in the United States
Habitat	Domesticated
Diet	Meat, fish, and some vegetables
Breeding	Around 6 kittens up to three times a year

American Curl

The curled ears of this breed are the result of a genetic mutation. In 1981 a long-haired kitten showing this characteristic was born in the California home of Joe and Grace Ruga. They later noticed that two of this cat's own kittens carried the mutation, suggesting that it arises from a dominant gene—and this allowed them to develop the breed. All kittens in an American curl litter are born with apparently normal ears, and about half will go on to develop curled ears. Both long- and short-haired American curls are available, in all colors and coat patterns. The body and face shape are midway between Oriental breeds and the short, compact shape exemplified by the British shorthair. These cats make intelligent and playful pets, with no health issues arising from their unusual ears.

Scientific name	*Felis catus*
Family	Felidae
Size	2.2–4.5kg (5–10lb)
Distribution	Originated in the United States
Habitat	Domesticated
Diet	Meat, fish, and some vegetables
Breeding	Around 4 kittens up to three times a year

Chartreux

One story holds that this ancient breed was developed at the monastery of Grande Chartreuse, the center of the Carthusian order, near Grenoble in France. Another story tells us that the Chartreux is descended from Syrian wildcats brought to France by returning Crusaders, many of whom joined the Carthusian order. Whatever its origins, the Chartreux is highly prized for its unique fur and affectionate nature. The only coat color for the breed is solid blue-gray, with silver tipping to the hairs giving an attractive sheen. The Chartreux is also known for its "smile," an appearance given by the structure of its head and muzzle. A large and muscular cat, the Chartreux makes a first-class hunter. However, in the home, this cat is quiet, playful, and good with children.

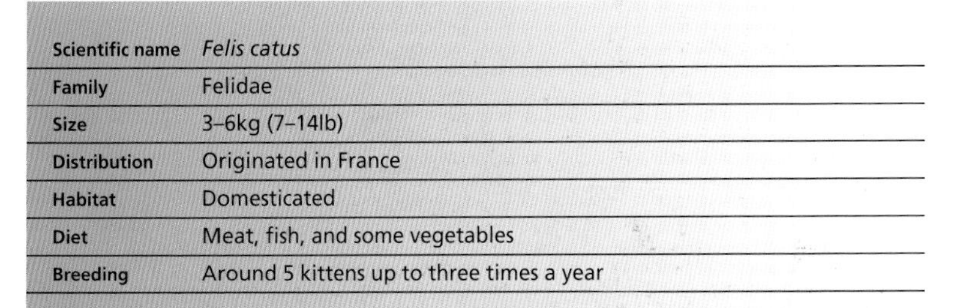

Scientific name	*Felis catus*
Family	Felidae
Size	3–6kg (7–14lb)
Distribution	Originated in France
Habitat	Domesticated
Diet	Meat, fish, and some vegetables
Breeding	Around 5 kittens up to three times a year

Bengal

Bengals are a hybrid developed by crossbreeding domestic cats with wild Asian leopard cats. A leopard cat is the size of a domestic cat and lives in forests, hunting for small mammals and birds. Over several generations, the wildcat was bred with domestic breeds such as American shorthairs, Egyptian maus, and Abyssinians. The Bengal has the temperament of a domestic cat with the markings and body structure of the wildcat. Bengals have either spotted or marbled coat patterns. A typical face shows horizontal stripes, known as "mascara," extending from the eye to the back of the neck. Separated from their wildcat ancestor by four generations, Bengals are well socialized and happy to join in with the bustle of the average home. They love children and are even happy to play with family dogs.

Scientific name	*Felis catus*
Family	Felidae
Size	3–6.8kg (7–15lb)
Distribution	Originated in the United States; licenses may be required in some countries
Habitat	Domesticated
Diet	Meat, fish, and some vegetables
Breeding	Around 4 kittens up to three times a year

Japanese Bobtail

With a long history in its native Japan, the Japanese bobtail is relatively rare in the rest of the world. The eponymous bobbed tail—which is about 10 centimetres (4 inches) long—is caused by a naturally occurring recessive gene, which means it can be passed on if it is carried by both parents. The tail is usually kept curled close to the body, but may be carried upright when the cat is walking. Since the hair of the tail is longer than that on the rest of the body, it resembles a pom-pom. Japanese bobtails are accepted in almost any color, but the *mi-ke* coloration of white, black, and brown, and also bicolors, are particularly popular in Japan. These are sweet-tempered and talkative cats that adore human interaction.

Scientific name	*Felis catus*
Family	Felidae
Size	2.2–4.5kg (5–10lb)
Distribution	Originated in Japan
Habitat	Domesticated
Diet	Meat, fish, and some vegetables
Breeding	Around 4 kittens up to three times a year

Russian Blue

Also known as an Archangel blue, the Russian blue is a naturally occurring breed originating in Arkhangelsk (Archangel), Russia. This popular cat is renowned for its intelligence and unique silver-blue coat. Some breeders claim that the coat is hypoallergenic, although no cat can be entirely free from allergens. This wonderfully soft-to-touch coat is called a "double coat," with the undercoat being downy and equal in length to the guard hairs, which are blue with silver tips. It is these tips that give the coat its shimmering appearance. With its vivid emerald-green eyes, the Russian blue is remarkably attractive. It is quick to respond to its owner's emotions, bonds closely with its loved ones, and enjoys playing with other pets and children.

Scientific name	*Felis catus*
Family	Felidae
Size	3–6kg (6.5–13lb)
Distribution	Originated in Russia
Habitat	Domesticated
Diet	Meat, fish, and some vegetables
Breeding	6 kittens up to three times a year

Korat

The Korat is known as the "good-luck cat" in its native Thailand and is often given to newlyweds as a gift. The first known mention of the breed is in a Thai book of cat poems known as *Tamra Maew,* which probably dates back several hundred years. In the manuscript, the breed is described as having "smooth hairs with tips like clouds and roots like silver." Modern breeders have been careful to preserve the cat's distinctive appearance. The Korat has short, fine fur in a slate blue-gray, and vivid green eyes. The cat's delicate head is distinctly heart-shaped. The breed makes a playful and active pet, forming strong bonds with its owners. Well known for its soft voice, the Korat has a tendency to be shy with strangers.

Scientific name	*Felis catus*
Family	Felidae
Size	2.7–5kg (6–11lb)
Distribution	Originated in Thailand
Habitat	Domesticated
Diet	Meat, fish, and some vegetables
Breeding	Around 5 kittens up to three times a year

Burmese

This cat's friendly and playful nature, coupled with its elegant shape, make it a popular breed in both Europe and the United States. The British Burmese must lie midway between the lithe, "foreign" build of the Siamese and the stocky shape of cats such as the British shorthair. In the United States, the Burmese is slightly shorter and stockier than its British counterpart, with a fuller face. The coats of all Burmese, regardless of provenance, should be short and glossy, fading to a paler shade on the underparts. Widely recognized colors are sable (known as brown in the United Kingdom), blue, champagne (chocolate), and platinum (lilac). Some cat associations also recognize tortoiseshell varieties and colors such as red, cinnamon, and cream.

Scientific name	*Felis catus*
Family	Felidae
Size	3.6–5.4kg (8–12lb)
Distribution	Originated in Thailand and Myanmar (Burma)
Habitat	Domesticated
Diet	Meat, fish, and some vegetables
Breeding	Around 6 kittens up to three times a year

Bombay

The Bombay breed was developed in the United States in the late 1950s by crossing a black American shorthair with a sable Burmese, with the aim of creating a domestic cat resembling a miniature black panther. The resulting cat has striking gold or copper eyes, a short black coat with a strong sheen, and a muscular body shape closely resembling the sleek Burmese. In the United Kingdom, the term "Bombay" is used to describe a cat of the Burmese type with short black hair and copper to greenish eyes. The coats of both breeds require little attention. Both the British and American cats resemble the Burmese in personality, making playful, good-natured, and affectionate pets. They bask in human attention and are usually tolerant of children and dogs.

Scientific name	*Felis catus*
Family	Felidae
Size	3.6–5.4kg (8–12lb)
Distribution	Originated in the United States or United Kingdom
Habitat	Domesticated
Diet	Meat, fish, and some vegetables
Breeding	Around 6 kittens up to three times a year

Singapura

A great deal of controversy has surrounded the Singapura's origins. The original claim was that the cat is descended from street cats found in Singapore in 1975 and then imported into the United States. It was later discovered that some of the foundation cats were Burmese–Abyssinian crosses descended from cats born in the United States. Whatever this attractive cat's true origins, it is renowned for its small and delicate build, crowned by large, almond-shaped eyes and deep-cupped ears. Only one color of Singapura is officially recognized: the sepia agouti. The cat is sable brown with ivory ticking and an ivory underside, chest, and chin. Each ticked hair has alternating bands of dark and light, with a dark tip. Inquisitive and playful, the Singapura enjoys having plenty of human interaction.

Scientific name	*Felis catus*
Family	Felidae
Size	2.2–3.6kg (5–8lb)
Distribution	Originated in Singapore
Habitat	Domesticated
Diet	Meat, fish, and some vegetables
Breeding	Around 3 kittens up to three times a year

Abyssinian

It is believed that all Abyssinians descend from a kitten named Zula who was taken to Britain from Egypt after the 1868 British expedition to Abyssinia (modern-day Ethiopia). Zula was of "foreign" appearance, with a lithe body, wedge-shaped head, and almond-shaped eyes. Today, the European Abyssinian displays a more foreign appearance than the shorter-faced American Abyssinian. Abyssinians have fine, short hair. Each hair is ticked with darker-colored bands, lightest at the root and darkening toward the tip. The most common colors are ruddy, known as "usual" in the United Kingdom (reddish-brown with black ticking); fawn (gray with fawn ticking); sorrel (apricot with chocolate ticking); and blue (cream with blue ticking). Rarer colors are silver, lilac, cream, chocolate, and tortoiseshell. The Abyssinian is known for being intelligent, willful, and slightly aloof.

Scientific name	*Felis catus*
Family	Felidae
Size	2.7–4.5 (6–10lb)
Distribution	Originated in the United Kingdom
Habitat	Domesticated
Diet	Meat, fish, and some vegetables
Breeding	Around 3 kittens up to three times a year

Ocicat

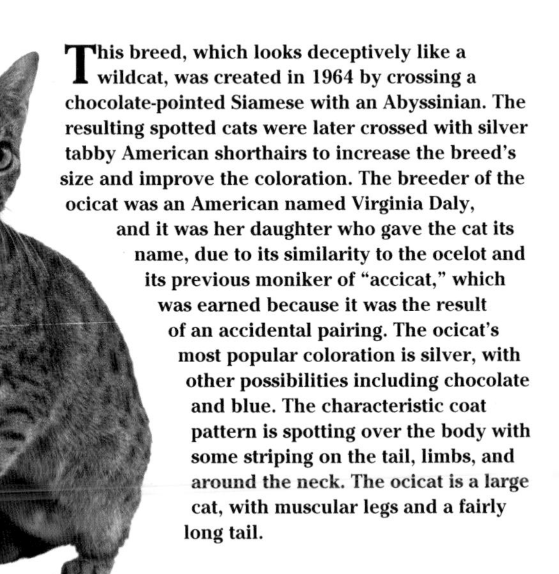

This breed, which looks deceptively like a wildcat, was created in 1964 by crossing a chocolate-pointed Siamese with an Abyssinian. The resulting spotted cats were later crossed with silver tabby American shorthairs to increase the breed's size and improve the coloration. The breeder of the ocicat was an American named Virginia Daly, and it was her daughter who gave the cat its name, due to its similarity to the ocelot and its previous moniker of "accicat," which was earned because it was the result of an accidental pairing. The ocicat's most popular coloration is silver, with other possibilities including chocolate and blue. The characteristic coat pattern is spotting over the body with some striping on the tail, limbs, and around the neck. The ocicat is a large cat, with muscular legs and a fairly long tail.

Scientific name	*Felis catus*
Family	Felidae
Size	3.6–8kg (8–18lb)
Distribution	Originated in the United States
Habitat	Domesticated
Diet	Meat, fish, and some vegetables
Breeding	About 5 kittens up to three times a year

California Spangled

This cat was the brainchild of California scriptwriter Paul Casey. After a trip to Tanzania in 1971, he was inspired to create a breed of domestic cat physically resembling the leopard, with the aim of drawing attention to the plight of endangered wildcats. The spangled was bred from a range of cats—including nonpedigree cats from Egypt and Asia, seal point Siamese, silver-spotted tabby Persians, and spotted Manx—to create its unique patterning and body shape. Like wildcats, the spangled has a long, lean body; large paws; and a medium-length tail. All cats have spotting over their back and sides, with stripes evident on the throat, limbs, and tail. Colors range from gold and silver to brown. Spangled cats are fit, sociable, and intelligent—and entirely domesticated.

Scientific name	*Felis catus*
Family	Felidae
Size	4–8kg (9–18lb)
Distribution	Originated in the United States
Habitat	Domesticated
Diet	Meat, fish, and some vegetables
Breeding	Around 4 kittens up to three times a year

Snowshoe

The well-named snowshoe was developed in the 1960s by an American breeder of Siamese cats named Dorothy Hinds-Daugherty. The snowshoe was created from a pairing of a Siamese and a bicolored American shorthair. The resulting cats display the characteristic dark points of a Siamese alongside striking white feet. These socks must reach to the ankles on the forelegs and to just below the hocks on the hind legs. Snowshoes appear with seal, chocolate, fawn, lynx, blue, or lilac points, along with blue eyes. Cats do not develop their full adult coloration for up to two years after birth. Snowshoes are lean and highly muscular. They are good-natured, playful, and fairly vocal. As a rule, they need plenty of attention and do not like to be left alone for too long.

Scientific name	*Felis catus*
Family	Felidae
Size	2.3–5.4kg (5–12lb)
Distribution	Originated in the United States
Habitat	Domesticated
Diet	Meat, fish, and some vegetables
Breeding	Around 5 kittens up to three times a year

Cornish Rex

The elegant Cornish rex has a uniquely silky and curly coat. Most cats have three types of hair: outer fur (or guard hairs), a middle layer called awn hair, and a short undercoat called down hair. Due to a natural genetic mutation, the Cornish rex has only the soft down hair. All cats of this breed are descended from a kitten born in Cornwall, England, in the 1950s. Despite claims that the Cornish rex is hypoallergenic, allergy sufferers may still have a reaction to the cat's dander and saliva. The cat's light coat means that it is best suited to indoor conditions. It will enjoy seeking out warm places, such as radiators and human laps and shoulders. This cat loves to play throughout its life and will engage in acrobatics, games of fetch, and much inquisitive mischievousness.

Scientific name	*Felis catus*
Family	Felidae
Size	2.5–4.5kg (5.5–10lb)
Distribution	Originated in Cornwall, England
Habitat	Domesticated
Diet	Meat, fish, and some vegetables
Breeding	Around 6 kittens up to three times a year

Devon Rex

When this breed was first developed, its soft, curly coat was thought to be caused by the same mutation as that of the Cornish rex, but the Devon rex has a distinctly different coat, with both down and guard hairs present. "Rex" refers to the breed's curly hair. The Devon rex was developed in the 1960s in Devon, England, from the offspring of a feral curly coated cat. Breeding with Siamese and Burmese has contributed to the cat's svelte appearance. The breed is often known as the "pixie cat" for its large, low-set ears and mischievous behavior. Affectionate and almost doglike in its eagerness to play, the Devon rex enjoys being at the center of a family. Its lack of shedding may make this breed less problematic for allergy sufferers, but no cat is wholly allergen-free.

Scientific name	*Felis catus*
Family	Felidae
Size	2.3–4.5kg (5–10lb)
Distribution	Originated in Devon, England
Habitat	Domesticated
Diet	Meat, fish, and some vegetables
Breeding	Around 5 kittens up to three times a year

Sphynx

The first successful breed of apparently hairless cats, the sphynx is descended from naturally mutated hairless cats that were discovered in the late 1970s. Hairless cats appear naturally about every fifteen years somewhere in the world. Although the sphynx seems hairless, the skin is actually covered in very soft hair like the fuzz on a peach. Its skin is the color that its fur would be, with all the usual cat marking patterns. Although some people with cat allergies react well to being around sphynx cats, other sufferers have been known to have strong reactions, since cat allergies are caused by cat saliva and dander rather than hair. Since the sphynx has no hair to keep it warm, it often cuddles up with its owners, and even enjoys sleeping under their bedcovers.

Scientific name	*Felis catus*
Family	Felidae
Size	3.4–5.4kg (7.5–12lb)
Distribution	Originated in the United States and Canada; rare
Habitat	Domesticated
Diet	Meat, fish, and some vegetables
Breeding	Around 4 kittens up to three times a year

Selkirk Rex

In 1987 a nonpedigree shorthair kitten with curly hair was discovered in Montana. The kitten also sported curly whiskers and frizzy ear tufts. This curly hair, which is the result of a genetic mutation, gives the cat its "rex" title. The kitten was later paired with a black Persian, and three of the resulting litter displayed curly hair, forming the foundation of the Selkirk rex breed. The breed has since been developed in both long- and short-haired types. In both cases, the coat is soft and woolly, with a high propensity for shedding. Unlike the Devon or Cornish rex, this breed is never recommended for anyone suffering from cat allergies. The Selkirk is solidly built, with a rounded head and wide eyes. The breed is known for being active and affectionate.

Scientific name	*Felis catus*
Family	Felidae
Size	3.6–5.4kg (8–12lb)
Distribution	Originated in the United States
Habitat	Domesticated
Diet	Meat, fish, and some vegetables
Breeding	Around 6 kittens up to three times a year

Siamese

The popular Siamese cat has a cream base coat with colored points on its muzzle, ears, lower legs, and tail. The pointed pattern results from a mutation in an enzyme involved in melanin production. The mutated enzyme is heat-sensitive: it fails to work at normal body temperature but becomes active in cooler areas of the skin, such as the cat's extremities. Siamese kittens are cream or white but develop visible points within a few months of birth. Siamese have bright blue, almond-shaped eyes. Many early Siamese cats had cross-eyes, but today this has largely been bred out. A Siamese cat's unusual loud cry is instantly recognizable and has often been compared to the sound of a baby crying. This breed is extremely dependent on humans and may bond closely with a particular individual.

Scientific name	*Felis catus*
Family	Felidae
Size	4–6.4kg (9–14lb)
Distribution	Originated in Southeast Asia
Habitat	Domesticated
Diet	Meat, fish, and some vegetables
Breeding	6 or more kittens up to three times a year

Tonkinese

A cross between the Siamese and Burmese cats, the Tonkinese was developed in Canada in the 1930s. A Tonkinese litter is likely to contain Siamese and Burmese as well as Tonkinese kittens. Many owners feel that the Tonkinese combines the best characteristics of both breeds. Tonkinese are lithe and muscular, displaying a modified wedge-shaped head with a square muzzle. With short, glossy fur, the Tonkinese displays three main coat patterns: solid, pointed, and mink. Solid coats come from Burmese ancestry, pointed coats are typically Siamese, while mink is a unique Tonkinese pattern, with subtly shaded points and a harmonizing body color. Colors include blue, lilac, cream, chocolate, and brown. "Tonks" resemble the Burmese in temperament, being less highly strung than the Siamese and with a less piercing voice. They make affectionate, curious, and playful pets.

Scientific name	*Felis catus*
Family	Felidae
Size	2.7–4.5kg (6–10lb)
Distribution	Originated in Canada
Habitat	Domesticated
Diet	Meat, fish, and some vegetables
Breeding	Around 4 kittens up to three times a year

Oriental Shorthair

The Oriental shorthair is a "nonpointed" member of the Siamese family. The Siamese cat was imported to Britain from Thailand in the nineteenth century in both solid and pointed colors (with a pale body and darker extremities). When the cats were bred, the pointed cats were eventually registered as Siamese while the others came to be known as foreign shorthairs or Oriental shorthairs. There are more than 300 different color and pattern combinations for the Oriental shorthair, including smoke, tabby, and bicolored. Just like the Siamese, Oriental shorthairs have emotional almond-shaped eyes, a wedge-shaped head with large ears, and an elegantly muscular body. This intelligent and social cat bonds closely with people—so closely, in fact, that some say they are doglike in personality. Shorthairs are inquisitive, demanding, and frequently vocal.

Scientific name	*Felis catus*
Family	Felidae
Size	4–6.4kg (9–14lb)
Distribution	United States, United Kingdom, Europe, and Thailand
Habitat	Domesticated
Diet	Meat, fish, and some vegetables
Breeding	6 or more kittens up to three times a year

Havana Brown

The Havana brown is related to the Oriental shorthair and Siamese cats. This short-haired cat with large, forward-tilting ears and expressive green eyes is renowned for its solid-color chocolate-brown coat. The origins of the breed's name may lie in the coat color's similarity to that of Havana cigars. The Havana is an intelligent cat that uses its paws uniquely, both to examine objects and to communicate with its owners. A typical greeting may be to stretch out and raise a paw. A highly people-oriented animal that does not like to be left alone for long periods, this cat will manage a relocation well because it is less interested in territory than are other breeds. Some owners say that their pet can be doglike in personality. The Havana has never been bred in large numbers and remains rare.

Scientific name	*Felis catus*
Family	Felidae
Size	3.6–4.5kg (8–10lb)
Distribution	Originated in England; rare
Habitat	Domesticated
Diet	Meat, fish, and some vegetables
Breeding	Around 6 kittens up to three times a year

Egyptian Mau

This ancient and relatively rare breed is the only naturally spotted domestic cat, and is said to be descended from African wildcats. This unusual cat is also the fastest domestic cat: its long hind legs and unique flap of skin extending from the flank to the back knee allow for greater agility and speed. Maus have been recorded running at over 48 km/h (30 mph). The mau's eyes are gooseberry green and its coat may be silver, bronze, or smoke. When stimulated, maus can chirp and chortle musically. A joyous mau may display a characteristic dance of moving its back legs up and down as if spraying to mark its territory, without actually releasing any urine. This intelligent and alert cat requires more effort than other breeds but will repay a loving and attentive owner with devotion.

Scientific name	*Felis catus*
Family	Felidae
Size	2.5–5kg (5.5–11lb)
Distribution	Mainly Egypt, United States, and Italy
Habitat	Domesticated
Diet	Meat, fish, and some vegetables
Breeding	Around 6 kittens up to three times a year

Nonpedigree

Nonpedigree cats are by far the most popular choice for cat owners. In fact, 97 percent of domestic cats worldwide do not belong to a registered breed. One good reason for choosing a "moggy" is that it can be obtained from an animal shelter, enabling you to do your bit to reduce the dramatic overpopulation of cats. Moggies have a life span similar to that of pedigrees, of fifteen years or more. All personalities, coat lengths, colors, and patterns occur, with bicolors and tabbies (pictured) being common. Short-haired nonpedigree cats vary widely in appearance depending on their parentage, but many have evolved into a similar form due to repeated matings between common types. Many short-haired moggies have a stocky body shape and a fairly rounded face. Most long-haired moggies have slightly less luxurious fur than their pedigreed counterparts.

Scientific name	*Felis catus*
Family	Felidae
Size	2.3–6.8kg (5–15lb)
Distribution	Worldwide
Habitat	Domesticated
Diet	Meat, fish, and some vegetables
Breeding	3–5 kittens up to three times a year

Persian

The ancestors of today's Persian cat arrived in Europe from Persia in the seventeenth century. In Great Britain, where the Persian is often known as the longhair, the breed was crossed with the Angora. More than 300 years of selective breeding have since made the Persian a great deal stockier and extensively shortened its muzzle. A highly foreshortened muzzle can cause breathing problems, so conscientious breeders will breed only cats with a more moderate head shape. Known for being a good-natured pet, the Persian has a long, thick coat that needs daily grooming to prevent matting. A wide range of colors and markings are accepted, including solid colors, bicolors, tortoiseshell, and tabby. Tipped varieties are known as Chinchillas, while pointed Persians are called Himalayans in the United States and color-pointed longhairs in Europe.

Scientific name	*Felis catus*
Family	Felidae
Size	4–6.4kg (9–14lb)
Distribution	Originated in Iran
Habitat	Domesticated
Diet	Meat, fish, and some vegetables
Breeding	Around 4 kittens up to three times a year

Birman

The Birman is believed to have originated in Myanmar (Burma), where it was a temple cat. Originally, a legend tells us, all the temple cats were white, but as a priest named Mun-Ha lay dying, a cat named Sinh leaped onto his head. While his paws remained white where he had touched the priest, his coat became golden like a statue; his ears, nose, tail, and legs became dark like the earth; and his eyes were turned to sapphire. Whatever its origins, the breed retains these characteristics to this day. A Birman has white "gloves" and an eggshell or golden body, while its markings are seal, chocolate, blue, red, lilac, or cream. Tabby and tortoiseshell variations occur. An intelligent and inquisitive cat, the Birman forms strong bonds with its owners.

Scientific name	*Felis catus*
Family	Felidae
Size	3.6–6.8kg (8–15lb)
Distribution	Originated in Myanmar (Burma)
Habitat	Domesticated
Diet	Meat, fish, and some vegetables
Breeding	Around 6 kittens up to three times a year

Scottish Fold Longhair

This breed is descended from a nonpedigree short-haired kitten with folded ears, born close to Coupar Angus in Scotland in 1951. Pedigree British shorthairs were later used to develop the breed's bloodline. The cat's ears, which are caused by a natural genetic mutation, are folded forward toward the front of the head. Combined with the large, round eyes, these ears give the cat a rather owl-like appearance. Scottish folds may be short-haired or long-haired, with the latter being known variously as Highland fold, Scottish fold longhair, and longhair fold, depending on the registering body. A Scottish fold may be nearly any color or pattern apart from pointed. Bicolored Scottish folds tend to be the most popular. The breed is known for its placid, affectionate, and playful nature, making it an ideal family pet.

Scientific name	*Felis catus*
Family	Felidae
Size	2.7–6kg (6–13lb)
Distribution	Originated in Scotland
Habitat	Domesticated
Diet	Meat, fish, and some vegetables
Breeding	Around 5 kittens up to three times a year

Turkish Angora

This naturally occurring breed originated in the Ankara region of central Turkey and is related to the Turkish Van, sharing its liking for water. These delicately structured cats often have a soft white coat, but nearly all colors and patterns are found. Eyes may be amber, green, or blue—or perhaps even one blue and one amber. As with most animals with a white coat and blue eyes, a blue-eyed Turkish Angora should have its hearing checked. However, deaf cats can lead a perfectly happy indoor life. Although this breed does not like to be held for long, it bonds closely with its owners and enjoys playing near them. It often attempts to take part in human conversations. Most unusually, the Turkish Angora may like to take a bath with its owner.

Scientific name	*Felis catus*
Family	Felidae
Size	2.5–6kg (5.5–13lb)
Distribution	Originated in central Turkey
Habitat	Domesticated
Diet	Meat, fish, and some vegetables
Breeding	Around 6 kittens up to three times a year

Turkish Van

The famous swimming cat evolved naturally in a mountainous region of eastern Turkey, in the Lake Van area. The region is extremely cold in winter and hot in summer, which is the cause of the breed's unusual coat. The water-resistant coat is thick and soft during winter, and a mature cat will even have a winter mane. In summer, the cat sheds this long hair to reveal a shorter, silky coat. The Van is largely white, with color limited to the head and tail. Eyes may be blue or amber, or one of each. Unlike the Turkish Angora, however, blue-eyed Vans are not usually deaf. A large cat, the Van likes to leap and play energetically. It may enjoy taking a swim in the bath. This friendly cat enjoys being in human company as much as possible.

Scientific name	*Felis catus*
Family	Felidae
Size	5–9kg (11–20lb)
Distribution	Originated in eastern Turkey
Habitat	Domesticated
Diet	Meat, fish, and some vegetables
Breeding	Around 5 kittens up to three times a year

Maine Coon

The Maine coon, named after the state of Maine, was the first breed of long-haired cat to originate in North America. It is believed that long-haired cats found their way to the early American colonies aboard visiting ships, where they were kept for rodent control. Surviving in the wild in the harsh winters of Maine, these cats developed into a large and hardy cat with a uniquely water-resistant, thick, long coat. Maine coons are accepted in all colors and patterns except lavender, chocolate, ticked tabby, and pointed. The most common coloration is brown tabby. Many tabbies sport an "M" marking on their forehead. Maine coons are highly dextrous, intelligent, and playful, frequently using their paws for all manner of tasks, from opening doors to flushing toilets.

Scientific name	*Felis catus*
Family	Felidae
Size	3–8kg (7–18lb)
Distribution	Originated in the United States
Habitat	Domesticated
Diet	Meat, fish, and some vegetables
Breeding	Around 5 kittens up to three times a year

Norwegian Forest

Known in its homeland as the *Norsk skaukatt*, this breed has been kept in Norway for centuries but has arrived in the United States and United Kingdom only in the last thirty years. The exact origins of the breed are unknown, but it is likely that the cat's hardy build and water-resistant, double-layered coat are simply natural adaptations to the harsh Norwegian climate. The Norwegian forest cat is large and muscular, with a triangular-shaped face. It has a long, bushy tail, while tufts of hair grow from its ears. A variety of coat colors and patterns are seen, including tabby, tortoiseshell, and bicolors. White fur, if present, is kept to the chest and paws. This active cat will do best in a home with plenty of outdoor space for hunting, ideally in open land.

Scientific name	*Felis catus*
Family	Felidae
Size	4–10kg (9–22lb)
Distribution	Originated in Norway
Habitat	Domesticated
Diet	Meat, fish, and some vegetables
Breeding	Around 4 kittens up to three times a year

Ragdoll

The ragdoll earns its name by its tendency to become limp when it is stroked, a trait that has been encouraged by selective breeding. This cat is known for its gentle temperament and strong attachment to its owners. Due to this sweet nature, a ragdoll should never be allowed outdoors without supervision. The breed was developed in the 1960s by Ann Baker, a Persian cat breeder in California, who crossed Persians with Birman and Burmese cats. Ragdoll coat colors are seal, chocolate, flame, blue, lilac, and cream. There are three different patterns: pointed, mitted (pointed, with white paws, chin, and stomach), and bicolor (with white on the limbs and stomach, and in an inverted V on the face). Lynx ragdolls, which are found in all three patterns, have stripes on their tail, face, and points.

Scientific name	*Felis catus*
Family	Felidae
Size	4.5–9kg (10–20lb)
Distribution	Originated in the United States
Habitat	Domesticated
Diet	Meat, fish, and some vegetables
Breeding	Around 5 kittens up to three times a year

Siberian Forest

This naturally occurring breed is believed to have been in existence in its Russian homeland for at least a thousand years. It is thought by some to be the ancestor of all the long-haired breeds. As with the Norwegian forest cat, the breed's long, thick, and water-resistant hair is likely to be a naturally occurring adaptation to the cat's cold habitat. As yet, Siberians have not been selectively bred to any great extent. Siberians have longer hair on the sides of the face and a ruff around the neck. They are often seen with tabby markings, with golden tabbies being highly prevalent. These cats are fairly short and stocky, with powerful legs and a broad, round head. Often described as having a doglike personality, Siberians are intelligent, playful, and loyal.

Scientific name	*Felis catus*
Family	Felidae
Size	4.5–9kg (10–20lb)
Distribution	Originated in Russia
Habitat	Domesticated
Diet	Meat, fish, and some vegetables
Breeding	Around 4 kittens up to three times a year

Tiffany

Also known as the Chantilly, the Tiffany is often described as a long-haired form of the popular Burmese. The long-haired gene was introduced by crossing Burmese with Persian longhairs. With a similar form to the Burmese, the Tiffany has a wedge-shaped head, widely spaced ears, and a muscular body. The cat's oval eyes range from amber to golden yellow. Its lustrous fur is semilong and appears in rich shades of chocolate, blue, cinnamon, silver, lilac, and fawn. Accepted patterns are solid, mackerel, ticked, and spotted tabby. The Tiffany is known for being loyal, bonding closely with its owners, and often chatting with them in characteristic "chirps." Owners who work full-time should consider getting a companion cat. Tiffanies are usually fond of children and may integrate well with other pets.

Scientific name	*Felis catus*
Family	Felidae
Size	2.7–4.5kg (6–10lb)
Distribution	Originated in the United States
Habitat	Domesticated
Diet	Meat, fish, and some vegetables
Breeding	Around 5 kittens up to three times a year

Oriental Longhair

Also sometimes known as the Javanese, foreign longhair, or Mandarin, the Oriental longhair emerged from a British breeding program in the 1970s aimed at re-creating the traditional Angora by crossing Abyssinians and Siamese. In fact, the breed was known as the British Angora until 2003, when it was renamed to prevent confusion with the Turkish Angora. The breed has the graceful and svelte build of an Oriental cat, with a wedge-shaped head and large, wide ears. Its semilong and silky coat is recognized in a wide range of colors and patterns, including tabby, tortoiseshell, smoke, tipped, and solid (in cinnamon, black, chocolate, blue, lilac, fawn, red, and cream). The Oriental longhair is known for its affectionate nature, and also displays the intelligence and inquisitiveness characteristic of Oriental cats.

Scientific name	*Felis catus*
Family	Felidae
Size	3–4kg (7–9lb)
Distribution	Originated in the United Kingdom
Habitat	Domesticated
Diet	Meat, fish, and some vegetables
Breeding	Around 5 kittens up to three times a year

American Bobtail

This is a relatively new and rare breed of cat that was recognized by the International Cat Association in 1989. The breed is known for its stubby tail, about a third to half the length of a normal cat's tail. This is the result of a naturally occurring dominant gene, which means that the mutation can appear even if the gene is inherited from just one parent. American bobtails can have coats and eyes of any color, and both long- and short-haired types are available. The preferred show bobtail has a "wild-looking" tabby coloration. Bobtails are stocky cats with a broad, wedge-shaped head. As a pet, the breed is friendly, active, and intelligent. Owners report that this is a cat skilled in making escapes, with closed doors being no barrier at all.

Scientific name	*Felis catus*
Family	Felidae
Size	3.2–7kg (7–15lb)
Distribution	Originated in the United States
Habitat	Domesticated
Diet	Meat, fish, and some vegetables
Breeding	Around 5 kittens up to three times a year

Banjo Catfish

The *Bunocephalus coracoideus* species is the most common banjo catfish found in the home aquarium. All fish in the Aspredinidae family are known as banjo catfish due to their large, flattened heads and slim tails, giving them the appearance of a banjo. Like all catfish, these fish are scaleless, but their skin is completely covered in the fibrous protein keratin and dotted with large tubercles, or lumps. Fish in the genus *Bunocephalus* are nocturnal. *Bunocephalus coracoideus* will do best in an aquarium with a sandy bottom, where it will bury itself for much of the day. At night, it can be seen scurrying across the bottom of the tank looking for food. This is a docile fish that is relatively easy to breed. The sexes can be distinguished because females are much fatter than males.

Scientific name	*Bunocephalus coracoideus*
Family	Aspredinidae
Size	11–15cm (4.5–6in)
Distribution	South America: Colombia to Brazil
Habitat	Freshwater: rivers with soft bottoms
Diet	Tablet food, frozen bloodworms, and tubifex worms
Breeding	4,000 eggs laid in sand

Hillstream Loach

There are more than 600 species of hillstream loaches in sixty genera, many of which are popular for freshwater aquariums. There are numerous loaches in the genus *Gastromyzon*, including the spiny-headed, striped, spotted, and saddleback hillstream loaches. For keeping a healthy shoal of these fish, a long and wide aquarium should have a gravel bed and feature driftwood and flow-loving plants. Highly oxygenated water, created with power heads and flow pumps, is needed to replicate the hillstream loach's natural habitat. Adequate lighting is necessary for algal growth. These fish can be fairly territorial and may get into disputes: skirmishes may involve one fish trying to "top," or cover, another fish. However, these loaches will generally tolerate others of their species and similar fish. Spiny-headed hillstream loaches are known for being more easygoing than other species.

Scientific name	*Gastromyzon ocellatus*
Family	Balitoridae
Size	4–6cm (1.6–2.4in)
Distribution	Southeast Asia: Sarawak, Borneo
Habitat	Freshwater: fast-flowing upland streams and rivers
Diet	Prepared flakes and pellets plus thawed frozen bloodworms, and blanched spinach and kale
Breeding	Not bred in aquariums

Bronze Corydoras

Belonging to the armored catfish family, this hardy and easy-to-breed fish is a popular aquarium species. It is not as beautiful as many freshwater aquarium fish, with its yellow or pink body, white belly, and blue-gray head and back. A dark orange patch on the head is visible from above. Yet this catfish is liked by many aquarium owners because it is constantly active and has the unusual habit of spending much of its time foraging along the bottom of the aquarium, disturbing clouds of waste material. The bronze corydoras prefers to be kept in a shoal and is well suited to a peaceful community aquarium. Since this fish is a bottom-feeder, it should be watched at feeding times to make sure that other fish are not consuming all the food before it gets a chance.

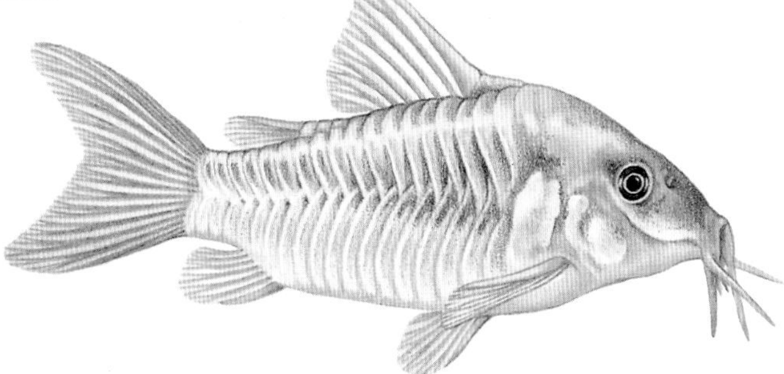

Scientific name	*Corydoras aeneus*
Family	Callichthyidae
Size	6–7cm (2.5–2.8in)
Distribution	South America: Colombia to northern Argentina
Habitat	Freshwater: shallow, muddy-bottomed waters
Diet	Flake food and live food such as worms and daphnia
Breeding	Up to 200 eggs laid on plant leaves

Cardinal Tetra

The popular cardinal tetra sports an iridescent blue line midway down its sides, with the body below this line being cardinal red. The closely related neon tetra, which is cared for in a similar manner, looks identical, but its red coloration is only on the back portion of its body. A local—and environmentally sustainable—industry of cardinal tetra harvesting takes place in Brazil on the banks of the Negro River. This fish prefers to be kept in a shoal of half a dozen or more individuals in an aquarium heated to 21–28°C (70–82°F). An ideal aquarium setup to promote breeding would include bogwood, live native plants, some floating plants to provide hiding places, and subdued lighting. The cardinal tetra will do well in an Amazon community aquarium with other peaceful fish.

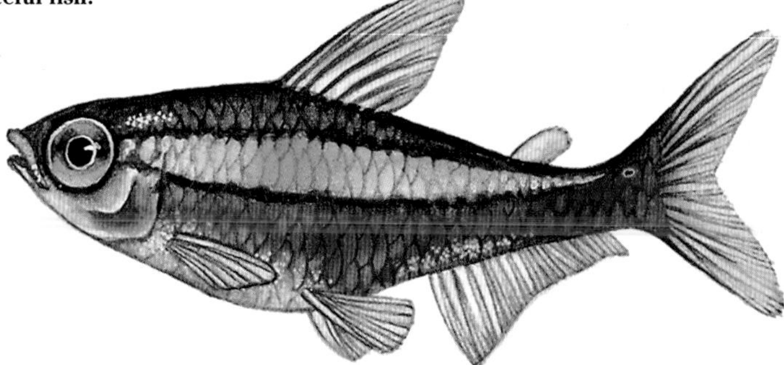

Scientific name	*Paracheirodon axelrodi*
Family	Characidae
Size	3cm (1.25in)
Distribution	South America: upper Orinoco and Negro rivers
Habitat	Freshwater: rivers deep in rain forests
Diet	Flake fish food plus small worms
Breeding	500 eggs; sensitivity to water conditions makes them difficult to breed

Redeye Piranha

The redeye piranha, also known as the black piranha, does not tolerate any tankmates and must be kept alone. It has razor-sharp teeth and powerful jaws, so care must be taken when carrying out tank maintenance! Piranhas tend to become more aggressive with age. Obviously, they are not a fish to be kept in a home with children. Some countries and states have made ownership of piranhas illegal. In the wild, these carnivorous fish live in loose shoals and are generally nonaggressive, but in the low-water season when fish become concentrated in pools, hungry piranhas can be a danger to any animal or human entering the water. The feeding of live fish and mammals to pet piranhas is controversial. A healthy diet might include live foods such as worms, fish foods, and frozen meat and fish.

Scientific name	*Serassalmus rhombeus*
Family	Characidae
Size	20–39cm (8–15.5in)
Distribution	South America: Amazon and Orinoco river basins and Brazilian coastal rivers; it is illegal to own piranhas in some countries and states
Habitat	Freshwater: rapids and deep zones of rivers
Diet	Insects, freeze-dried fish foods, frozen meat and fish
Breeding	Not bred in aquariums

Freshwater Angelfish

P*terophyllum scalare* is the freshwater angelfish most frequently kept in home aquariums, earning it its common name. Its disk-shaped body is very thin when viewed from the front, and it has long, flowing dorsal and anal fins. In addition to the standard silver coloration seen in the wild, many different colors and patterns have been developed in captivity, including gold, blue, black, marbled, zebra, and lace. In the wild, the freshwater angelfish preys on small fish and invertebrates by ambushing them. In the home aquarium, it should be kept with docile fish of a similar size to avoid fin-nipping and attacks. The tank should be planted with hardy species to provide plenty of hiding places. When well cared for, this fish can live for up to fifteen years in captivity.

Scientific name	*Pterophyllum scalare*
Family	Chiclidae
Size	7.5cm (3in)
Distribution	South America: Peru, Colombia, Brazil, and French Guiana
Habitat	Freshwater: swamps or flooded ground with dense vegetation
Diet	Flakes, frozen foods, and live foods
Breeding	Up to 1,200 eggs laid on a leaf, rock, or aquarium side

Lifalili Jewel Cichlid

When spawning, this cichlid is blood red, with iridescent spots of yellow to turquoise. At other times, the fish is a less vibrant red, with an olive back. Males are less colorful than females. A stripe runs through the eye and there is a dark spot on the gill cover and the midsection. An aggressive and territorial fish, the jewel cichlid is not suited to the usual community tank; it may be safest to keep it in a single-species aquarium unless you are sure that tankmates are suitable, like the catfish in the genus *Synodontis* known as squeakers and Congo tetra. These cichlids require a well-oxygenated and spacious aquarium with plenty of stones and wood to hide around. They like to dig, so any plants will need to be strongly rooted.

Scientific name	*Hemichromis lifalili*
Family	Cichlidae
Size	10–12.5cm (4–5in)
Distribution	Central Africa: tributaries of the Zaire and Ubanghi rivers
Habitat	Freshwater: slow-moving creeks and tributaries
Diet	Live foods such as insect larvae and tubifex worms, plus pellets and flakes
Breeding	Up to 400 eggs laid on a rock or aquarium wall

Common Discus

The beautiful discus cichlids have a laterally compressed, button-shaped body, earning them their name. There are three subspecies of the common discus: blue, brown, and green. Many other color and pattern permutations have been developed. These fish are well known for the highly developed care of their young, with both parents taking part. The adults produce a secretion in their skin, on which the larvae feed. The common discus is shy and peaceful in an aquarium. It will do best with tankmates such as freshwater angelfish, cardinal tetras, and the bronze corydoras. Discuses are not hardy and are very sensitive to water temperature and quality. Their tank will need a gravel substrate and plenty of driftwood to provide hiding places and keep the water suitably acidic.

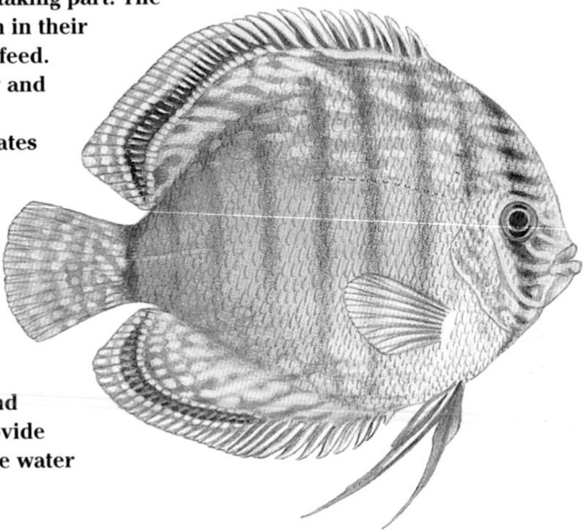

Scientific name	*Symphysodon aequifasciatus*
Family	Cichlidae
Size	20–25cm (8–10in)
Distribution	South America: Brazil, Colombia, and Peru
Habitat	Freshwater: deep rocky areas of rivers among crevices
Diet	Live black worms, bloodworms, brine shrimp, and mosquito larvae, plus some granules
Breeding	100–200 eggs laid on a leaf, rock, or aquarium wall

Texas Cichlid

The Texas cichlid, also known as the Rio Grande perch, has a grayish body and blue scales, with a dark spot at the body's center and another on the tail. A mature male has a large hump on its head. The most northerly of the cichlids, this fish prefers a large tank kept at a temperature of 20–33°C (68–91°F). A sandy substrate is needed, as this cichlid enjoys a good dig, sometimes uprooting plants. This is a territorial fish and requires an aquarium with well-defined territories and hiding places created by rocks and wood. It can usually be kept with large aquarium species such as silver dollars and oscars, as well as other cichlids. If kept in the right conditions, this cichlid can live for up to fifteen years.

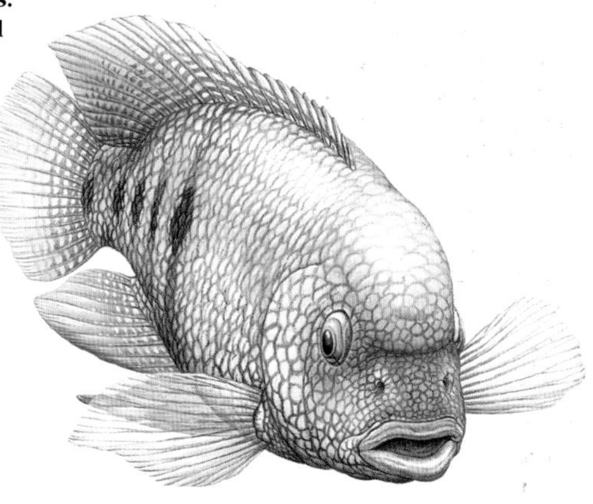

Scientific name	*Herichthys cyanoguttatus*
Family	Cichlidae
Size	12–30cm (5–12in)
Distribution	Lower Rio Grande system in Texas and northern Mexico
Habitat	Freshwater: sandy bottoms of deep rivers and lakes
Diet	Pellets, flakes, and live and frozen mosquito larvae and tubifex worms, spinach, and lettuce
Breeding	Up to 1,000 eggs laid on a flat surface

Goldfish

A domesticated member of the carp family, the goldfish was first kept in China during the Tang dynasty (618–907). It was introduced to Europe in the seventeenth century and to the United States by 1850. Selective breeding has produced many different colorations, body shapes, and fin and eye configurations. Some goldfish types, such as the common goldfish (pictured), shubunkins, and comet, can be kept in a pond year-round in temperate and subtropical climates. Other types, particularly those that have been bred to look least like the original fish, are not hardy and must be kept in an unheated aquarium, at 20–24°C (68–75°F). Goldfish should never be kept in a small fishbowl, as such an environment can cause stunting, deoxygenation, and ammonia or nitrite poisoning.

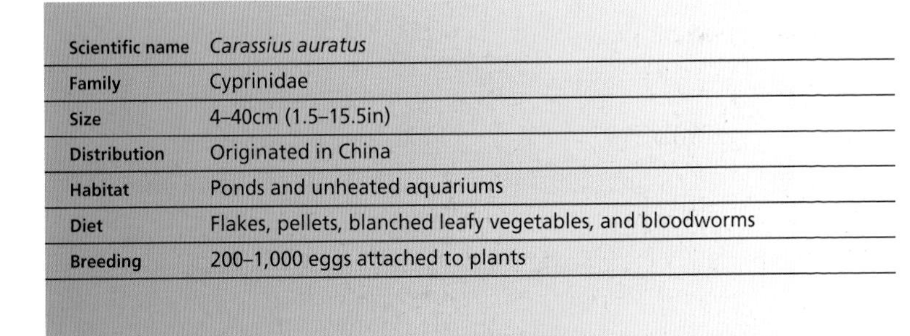

Scientific name	*Carassius auratus*
Family	Cyprinidae
Size	4–40cm (1.5–15.5in)
Distribution	Originated in China
Habitat	Ponds and unheated aquariums
Diet	Flakes, pellets, blanched leafy vegetables, and bloodworms
Breeding	200–1,000 eggs attached to plants

Harlequin Rasbora

This popular freshwater fish is the most easily available of the rasboras. The common name refers to the triangular black patch on the fish's orange-pink body, resembling the costume of the Harlequin. This fairly adaptable fish requires a well-cleaned aquarium of 21–28°C (70–82°F), with its ideal temperature for breeding being 28°C (82°F). The aquarium should be peppered with plants of the genera *Cryptocoryne* and *Aponogeton*, which provide ideal leaves for the depositing of eggs. Harlequins are happiest when kept in a shoal of at least half a dozen fish—and the spectacle of a darting shoal of harlequins can be a great pleasure. These peaceful fish do well in a community aquarium of fish with similar sizes and temperaments, such as other rasboras, barbs, danios, and catfish such as the bronze corydoras.

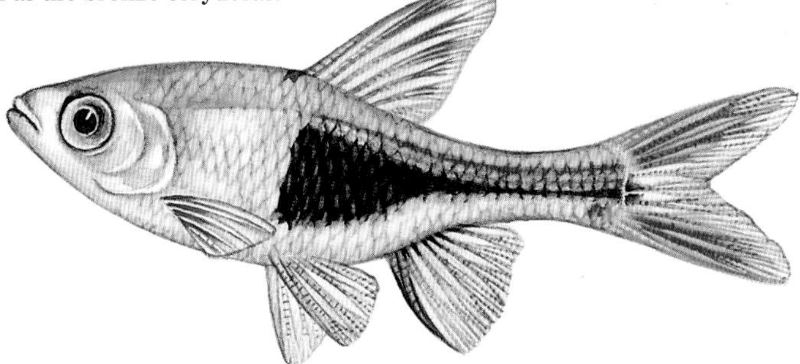

Scientific name	*Trigonostigma heteromorpha*
Family	Cyprinidae
Size	5cm (2in)
Distribution	Southeast Asia: Malaysia, Singapore, Sumatra, and southern Thailand
Habitat	Freshwater: streams in peat swamp forests
Diet	Prepared fish foods plus live foods such as mosquito larvae
Breeding	80–100 eggs deposited on the underside of plant leaves

Striped Raphael Catfish

Also known as the talking catfish, chocolate doradid, chocolate catfish, and thorny catfish, this widely available catfish is popular for its sociability in a community aquarium, living happily alongside other catfish and similarly docile species. This distinctly striped fish has rigid pectoral fin spines, which it sticks out even more firmly when stressed. For this reason, avoid using a net to catch this catfish because it can become tangled. In the wild, striped raphaels burrow in soft river bottoms during the day, and emerge to look for food such as mollusks, crustaceans, and organic waste at night. Young striped Raphaels have been observed cleaning larger fish, and it is thought that their stripe pattern may allow them to be recognized as a cleaner. In the home aquarium, these catfish prefer water heated to 24–30°C (75–86°F).

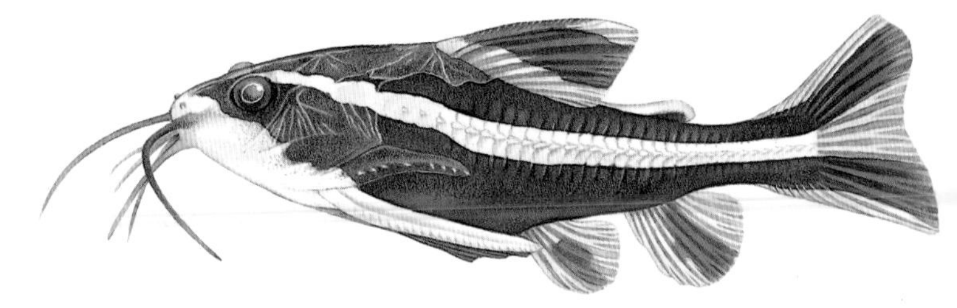

Scientific name	*Platydoras costatus*
Family	Doradidae
Size	20–24cm (8–9.5in)
Distribution	South America: French Guiana and Suriname
Habitat	Freshwater: rivers with soft bottoms
Diet	Flakes and tablets, plus bloodworms, earthworms, and tubifex worms
Breeding	Not bred in aquariums

Kissing Gourami

This fish earns its name with its protruding lips lined with horny teeth. These lips are used to rasp algae from stones and other surfaces in a "kissing" action. This kissing is also displayed by the frequently territorial males when challenging other fish. Sometimes the fish will "lip-lock" in aggressive wrestling competitions. Tankmates should generally be larger than the gourami and capable of fighting back. This algae-eating species is often used to control algae growth in an aquarium. The glass at the back of the tank should not be cleaned, so that the fish can feed on the algae growing there. There are two naturally occurring colors of the kissing gourami. The fish originating in Thailand are green, while fish from Java are pink with silvery scales and transparent fins.

Scientific name	*Helostoma temminckii*
Family	Helostomatidae
Size	20–30cm (8–12in)
Distribution	Southeast Asia: Thailand to Indonesia
Habitat	Freshwater: shallow, slow-moving, and thickly vegetated backwaters
Diet	Algae, flakes, vegetables, and frozen, freeze-dried, and small live foods
Breeding	Up to 1,000 eggs deposited in lettuce leaves that should be laid on the water surface

Royal Pleco

The royal pleco is patterned in pale gray with dark gray lines. Its eyes are red and its dorsal fins are edged with gold. This catfish is one of the few fish that can eat wood. Its aquarium should include bogwood, which it needs to chew on to stay healthy. The royal pleco does not swim well because its body armor, made of skin, makes it heavy and inflexible. However, it has a suckerlike mouth that it uses to hold onto wood and rocks in the rivers that are its home. It is friendly toward other fish in a community aquarium but does not like its territory to be invaded. The royal pleco should not be housed with fast-swimming fish because the other fish are likely to consume all the food before the royal pleco gets a chance.

Scientific name	*Panaque nigrolineatus*
Family	Loricariidae
Size	38–43cm (15–17in)
Distribution	South America: Amazon and Orinoco rivers
Habitat	Freshwater: fast-flowing rivers
Diet	Algae-based prepared food, vegetables, and bogwood
Breeding	Rarely bred in aquariums

Threadfin Rainbowfish

These rainbowfish are known for their beautiful trailing fins. The sexes have distinct fin shapes and coloration—and an individual's colors will vary depending on lighting, diet, and rank in the school. The male's body color is silver, with the top reflecting blue and the bottom orange. The fan-shaped first dorsal fin is a combination of black, yellow, red, and orange. The tail may be blue with red tips. Females are a treacly peach color with green lights and transparent fins. Threadfin rainbowfish are best kept in a group, with twice as many females as males. Although they can be housed in a peaceful community aquarium, they will breed and thrive most when in a single-species aquarium. Placing bogwood in the water will help create the acidic and swampy conditions that these fish enjoy.

Scientific name	*Iriatherina werneri*
Family	Melanotaeniidae
Size	3–5cm (1.2–2in)
Distribution	Australia, Indonesia, and Papua New Guinea
Habitat	Freshwater: swamps and thickly vegetated rivers
Diet	Small live foods such as baby brine shrimp and crushed flakes
Breeding	200 eggs scattered among plants

Siamese Fighting Fish

Also known as the betta, the Siamese fighting fish is renowned for its gorgeous coloration and flowing fins. In fact, these features have been achieved through selective breeding in captivity. In the wild, the betta has fairly short fins and is a dull green and brown unless it is excited, when it displays various strong colors. Varieties include the delta, veiltail, and halfmoon (named after their tail shapes), and the extra-large giant betta. Colors range from red, turquoise, or yellow to metallic shades. The betta is aggressive, so no more than one male should be housed in a tank, and a male and female should be placed together only if breeding. A group of females may coexist happily if they are able to establish a pecking order.

Scientific name	*Betta splendens*
Family	Osphronemidae
Size	5–7.5cm (2–3in)
Distribution	Southeast Asia: Mekong River basin
Habitat	Freshwater: densely vegetated, slow-flowing, and standing water
Diet	Pellets, bloodworms, brine shrimp, and finely chopped vegetables
Breeding	100–300 eggs deposited in bubble nests

Honey Gourami

The fish labeled "honey gourami" in pet stores is often not *Trichogaster chuna* but a colored variety of the thick-lipped gourami. This is because the honey gourami looks deceptively drab, unless the male is in his breeding colors, when his grayish-rust color changes to a spectacular red body, yellow top, and dark blue bottom. These shy fish need to be kept in a peaceful community tank, planted to provide plenty of hiding places. Males of this species should not be housed together unless the tank is roomy enough for them to establish separate territories. It is best to keep the sexes separated until spawning, to prevent an overzealous male from harassing a female. After spawning, the male builds a nest of bubbles, held together with saliva and plants, in which to house the eggs.

Scientific name	*Trichogaster chuna*
Family	Osphronemidae
Size	4–6cm (1.5–2.5in)
Distribution	India and Bangladesh
Habitat	Freshwater: thickly vegetated rivers and lakes
Diet	Flakes, freeze-dried and frozen foods, worms, and small vegetable tablets
Breeding	100–500 eggs deposited in bubble nests

Glass Catfish

Since this fish has no scales and no body pigment, it is transparent, earning it the common name of ghost fish. If you look closely enough, the fish's heart and guts are visible. In the clear streams and rivers that are this catfish's home, its transparency acts as camouflage. In the home aquarium, these shy fish prefer to be kept in a shoal of at least six individuals and like to hide in the shadows of plants and around rocks and logs. Glass catfish are highly sensitive to the pH balance and quality of aquarium water. A power head or pump should be used to create a strong current for them to swim against. They will do well in a community aquarium with other docile schooling fish, including the bronze corydoras, cardinal tetra, harlequin rasbora, and guppy.

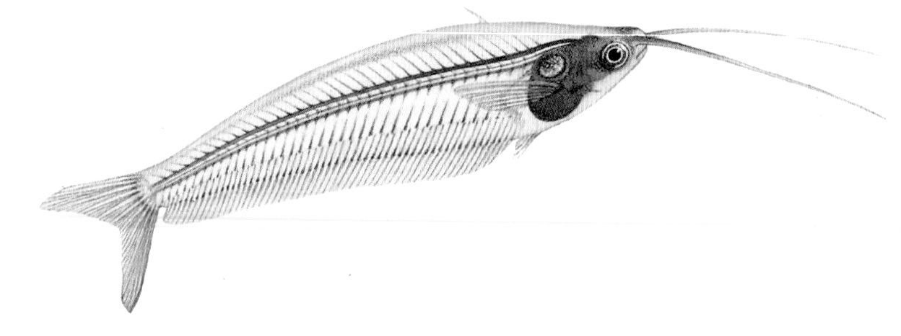

Scientific name	*Kryptopterus bicirrhis*
Family	Siluridae
Size	10–15cm (4–6in)
Distribution	Southeast Asia: Sumatra, Java, and Borneo
Habitat	Freshwater: fast-flowing, clear streams and rivers
Diet	Mosquito larvae, small worms, and brine shrimp; can be weaned to flake food
Breeding	Rarely bred in aquariums

Glass Knife Fish

As its name suggests, this elongated fish is transparent, although it has a brown tinge. Children may particularly enjoy watching this fish, as its backbone is visible through its body. Sometimes the fish's flanks have a green iridescence. The glass knife fish emits electrical signals during spawning. These "songs," which can be heard with the help of electrodes and an amplifier, are thought to help the fish find a mate. The glass knife fish is happiest in a group, which will order itself with a dominant fish and followers. It is best kept with other peaceful species too large to swallow. The relatively large aquarium should be heavily planted, with plenty of hiding places, as well as floating plants to dim the light. The water should be kept at 23–28°C (73–82°F).

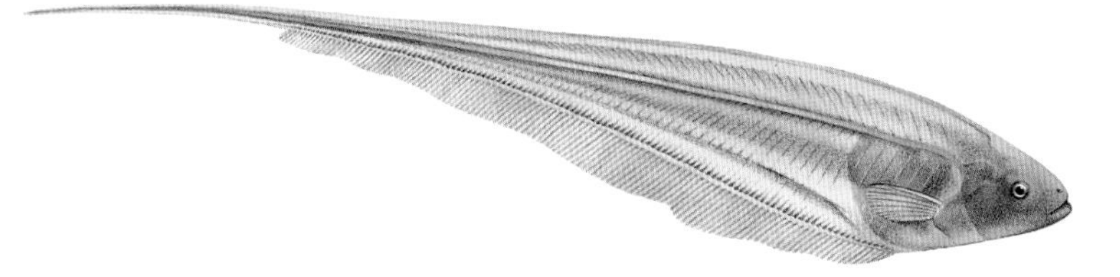

Scientific name	*Eigenmannia virescens*
Family	Sternopygidae
Size	22–45cm (8.5–17.5in)
Distribution	South America: east of the Andes, Colombia to northern Argentina
Habitat	Freshwater: underneath floating islands of vegetation in floodplains
Diet	Small fish, crustaceans, earthworms, mosquito larvae, tablets, and perhaps flakes
Breeding	100–200 eggs laid on roots of floating plants

American Flagfish

This patriotic American killifish species bears a fair resemblance to the Stars and Stripes flag. The male's body displays a large blue-black dot with alternating red and blue-black stripes. Females are much less colorful and distinct in their pattern. Females also sport a false eyespot in the center of their side, and another in the rear base of the dorsal fin. American flagfish are suitable for a brackish aquarium, kept at a cooler temperature of 19–26°C (66–80°F) and planted to provide hiding places. This hardy and adaptable species is nevertheless shy, so tankmates should be chosen accordingly. The flagfish is territorial, so each fish should be allowed enough space to set up a home patch. This species is a great algae eater, making particularly quick work of air algae.

Scientific name	*Jordanella floridae*
Family	Cyprinodontidae
Size	5–6cm (2–2.4in)
Distribution	North to Central America: Florida to the Yucatán
Habitat	Brackish water: still and slow-moving marshes, swamps, and lakes
Diet	Algae, green leafy vegetables, small invertebrates, and flake food
Breeding	100 eggs laid in a nest built by the male

Guppy

The guppy is perhaps the most popular freshwater or brackish aquarium fish of all. It is named after the nineteenth-century British naturalist who first described it, Robert John Lechmere Guppy. Male guppies have spots or stripes ranging from yellow through orange to red, and blue to purple or black. Females have gray bodies and brighter tails. Populations from different regions will display different colors and patterns. The domestic guppy has often been bred for its color, but as a result it can be extremely weak and prone to not surviving changes. Guppies live in complex social networks, with individuals remembering their relationships to other fish. Guppies are usually tolerant, but males can nip other males or different, enticingly finned species that enter their space. Guppies are known for being prolific breeders in both freshwater and marine aquariums.

Scientific name	*Poecilia reticulata*
Family	Poeciliidae
Size	2.5–6cm (1–2.5in)
Distribution	Caribbean islands and South America: Barbados to Brazil
Habitat	Fresh to brackish water: quiet, vegetated water
Diet	Baby brine shrimp, white worms, and flakes
Breeding	2–100 live young

Spotted Scat

The spotted scat is found in two colors, green and red. The green scat has a metallic silver sheen with a green cast. The red scat is similarly metallic, with a reddish cast. As its name suggests, there are black spots all across the scat's body. This beautiful fish is best kept in a large brackish aquarium when mature, with about 3–4 teaspoons of salt per 13 litres (2.9 gallons) of water. Scats are spawned in freshwater, but prefer an increasingly high salt content in their water as they age. If given enough space, the spotted scat is docile, but it can harass other species. The whole scat family are scavengers, feeding on algae and feces, among other foods, giving the family its name, Scatophagidae, from the Greek *skatos*, meaning "feces," and *phagein*, meaning "eat."

Scientific name	*Scatophagus argus*
Family	Scatophagidae
Size	30–35cm (12–14in)
Distribution	Indian and Pacific oceans: Kuwait to Japan and Australia
Habitat	Brackish, fresh, and salt water: river mouths to coastal waters
Diet	Frozen foods, flakes, aquarium plants, vegetables, and small live foods
Breeding	Never bred in aquariums

Targetfish

This brackish aquarium fish has curved bands running horizontally down its sides, making it resemble an archery target when viewed from above. It is silver with a black-banded tail. The targetfish should be mixed only with other large brackish fish, as it will be predatory toward smaller species. It should be kept in a school of at least four individuals, with twice as many females as males. The aquarium will need to be at least 300 litres (66 gallons) and kept at 24–28°C (75–82°F). The aquarium bottom should be covered with sand. In the wild, spawning takes place in salt water, before the juveniles swim to freshwater tidal pools and rivers, so these fish can be housed in a variety of water types, although brackish is preferred.

Scientific name	*Terapon jarbua*
Family	Terapontidae
Size	30–35cm (12–14in)
Distribution	Indian and western Pacific oceans
Habitat	Brackish, fresh, and salt water: estuaries and mangrove swamps to oceans
Diet	Frozen and prepared foods such as meaty pellets, shrimps, and clams; occasional live foods
Breeding	Rarely bred in aquariums

Achilles Tang

The Achilles tang is black with orange and white coloration along its fins and tail. As the fish matures, an orange teardrop shape develops on the back portion of the body, terminating in a sharp spine. These spines earned the Acanthuridae family its name, which means "thorn tail." The Achilles tang's small mouth has a single row of teeth for grazing on algae. Unfortunately, this surgeonfish has a low survival rate in captivity unless kept by a seasoned aquarist experienced in keeping other high-maintenance reef fish. The Achilles makes a peaceful tankmate if kept with fish such as clown fish, gobies, and butterfly fish, but may become aggressive if kept with other surgeonfish. Tangs require a large tank of at least 540 litres (120 gallons), as they need plenty of swimming room.

Scientific name	*Acanthurus achilles*
Family	Acanthuridae
Size	20–25cm (8–10in)
Distribution	Tropical Pacific Ocean
Habitat	Marine: seaward coral reefs
Diet	Benthic algae; brine and mysis shrimp
Breeding	Rarely bred in aquariums

Sohal Surgeonfish

With its blue and white horizontal stripes and dramatically flattened body, this surgeonfish makes a striking addition to a marine aquarium. The sohal surgeonfish sports a scalpel-like spine along the base of its tail on both sides. When threatened, it flicks the spines at its aggressor, sometimes causing severe injuries. This is one of the most aggressive tangs, but as long as only a single surgeonfish is housed in an aquarium of at least 540 litres (120 gallons), it can live for up to fifteen years in captivity. The sohal surgeonfish is a constant grazer and will snack on marine algae and nip at clam mantles and soft stony corals in the aquarium. When choosing a specimen to bring home, look for a fish that is swimming busily, breathing constantly, and looking alert and territorial.

Scientific name	*Acanthurus sohal*
Family	Acanthuridae
Size	Up to 40cm (16in); usually smaller in captivity
Distribution	Red Sea
Habitat	Marine: coral reefs, usually on steep slopes
Diet	Algae; brine and mysis shrimp
Breeding	Rarely bred in aquariums

Clown Triggerfish

This large fish, often found near reefs, displays a fantastically inventive coloration. The underside of its body is covered in large white spots on a dark background, while the dorsal, or upper, region has small black spots on a yellow background. This is an ideal camouflage for reef living. From below, the white spots resemble the light-dappled water surface. From above, the fish blends in with the corals among which it lives. From the side, the fish's silhouette is broken up, making it harder to spot. The species' startling, bright yellow mouth may be used to frighten away potential predators. The clown triggerfish requires a roomy aquarium so that it has plenty of swimming space. Tankmates should be chosen carefully because this fish can show aggression to other species and will prey on small fish and invertebrates.

Scientific name	*Balistoides conspicillum*
Family	Balistidae
Size	Up to 50cm (20in)
Distribution	Red Sea and tropical Indian and Pacific oceans
Habitat	Marine: clear coral reefs
Diet	Flake, squid, shrimp, chopped-up fish, and algae
Breeding	Rarely bred in aquariums

Mandarin Dragonet

The gorgeous mandarin dragonet gets its name from its vivid coloration, said to resemble the robes of a Chinese mandarin. Its body is covered in mazelike patterns of greens, blues, and yellows. The mandarin belongs to the dragonet, or "little dragon," family, characterized by long and scaleless bodies, with elongated and showy fins. Dragonets have large mouths and eyes, and fan-shaped tail fins. Despite their popularity for marine aquariums, mandarins are difficult to keep, as they are very picky feeders. Some fish never acclimatize to aquarium food. They are best kept in a mature reef tank with plenty of live rock and a self-sustaining population of copepods and other small crustaceans. Mandarins need a deep, sandy substrate, as they will often bury themselves at night or when startled.

Scientific name	*Pterosynchiropus splendidus*
Family	Callyonimidae
Size	5–6cm (2–2.5in)
Distribution	Western Pacific Ocean: Ryukyu Islands to Australia
Habitat	Marine: inshore or lagoon coral reefs
Diet	Tank population of copepods and small crustaceans, plus some frozen and prepared foods
Breeding	Rarely bred in aquariums

Copperband Butterfly Fish

This butterfly fish species is easily identified by its striking yellow stripes and long snout. It has an ocellus, or eyespot, toward the back of the dorsal fin. Unfortunately, this gorgeous fish is quite fragile and hard to keep, so it is not recommended for newcomers to marine aquariums. The copperband does not respond well to shipping and will need to be quarantined. As this species can be rather aggressive, it will do best in a single-species tank with only one male, or in a very large community tank. A reef fish, this butterfly fish will pick at corals kept in its tank if it is allowed to get hungry. The answer is to keep it well fed, enticing it with favorite live foods, such as clams offered in a half-shell at the bottom of the tank.

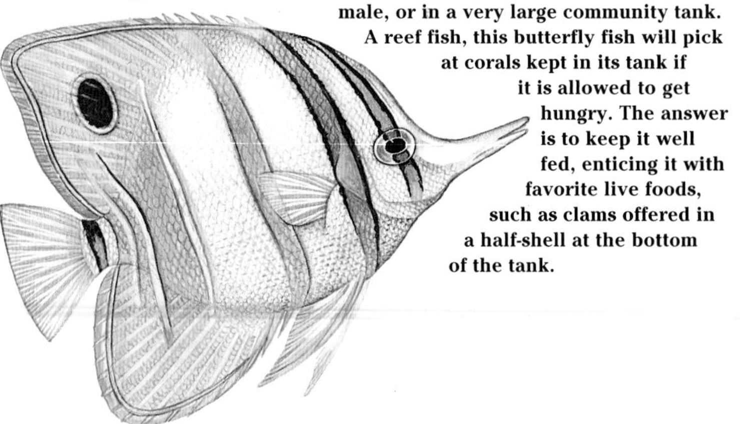

Scientific name	*Chelmon rostratus*
Family	Chaetodontidae
Size	15–20cm (6–8in)
Distribution	Western Pacific Ocean: Andaman Sea to Australia
Habitat	Marine: coral reefs, rocky coasts, and estuaries
Diet	Small live foods such as black worms, clams, and brine shrimp
Breeding	Rarely bred in aquariums

Clown Coris

Also known as the clown wrasse, this solitary fish can grow very large. Since it is also highly sensitive and difficult to keep, it should not be purchased by an inexperienced aquarist. This fascinating fish looks very different in its juvenile and adult colors. The juvenile is white and orange with eyespots on its dorsal fin. The adult, which develops a bulging forehead, is dark green with a paler band around its midsection. This coris will require a large tank, filled with rocks to create suitable cavelike hiding places. In its natural habitat, the clown coris likes to bury itself in the sand at night. To prevent this fish from injuring itself in the aquarium, the floor should be covered with 7–12 centimetres (3–5 inches) of sand.

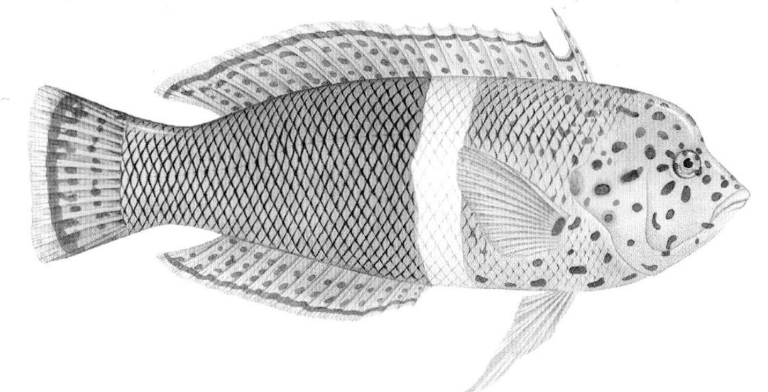

Scientific name	*Coris aygula*
Family	Labridae
Size	Up to 1.2m (4ft); usually much smaller in captivity
Distribution	Indian and western Pacific oceans
Habitat	Marine: seaward side of coral reefs close to sand
Diet	Fresh and prepared meaty foods such as shrimp and squid
Breeding	Rarely bred in captivity

Harlequin Tusk

This wrasse is suitable for a fish-only tank of moderately aggressive marine species. Larger angelfishes, wrasses, tangs, groupers, and puffers make suitable tankmates. Only one harlequin tusk should be kept in any aquarium. This fish owes its name to its protruding blue teeth, which are used for crushing crustaceans. The harlequin displays wide vertical orange and white bands over its head and most of its body. Its tail is yellow, and the rear of its body may be purple-blue. The wrasses are one of the few fish that bury themselves in the sand when sleeping or threatened. In the aquarium, most will sleep on the bottom of the tank in a sheltered rocky area. The harlequin tusk is fairly shy in its habits, but will grow bolder with age.

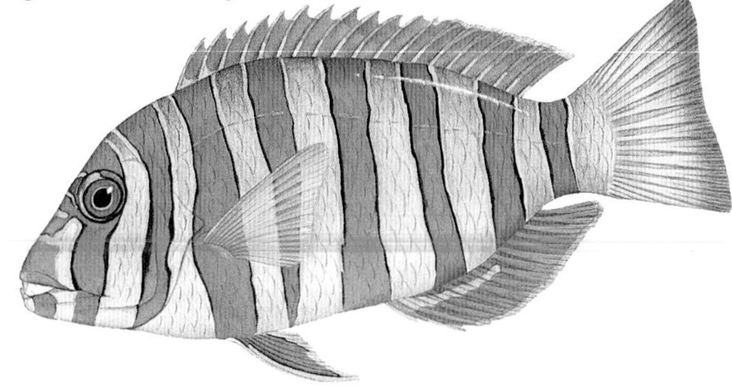

Scientific name	*Lienardella fasciata*
Family	Labridae
Size	20–25cm (8–10in)
Distribution	Indian and western Pacific oceans
Habitat	Marine: coral reefs
Diet	Small fish, crustaceans, and shellfish
Breeding	Not bred in aquariums

Red Sea Eightline Flasher

Ideal for a reef aquarium of similar-sized and nonaggressive fish, the Red Sea eightline flasher is a brilliantly colored wrasse. Its body is red and sports eight thin horizontal stripes, while its tail and fins are yellow. Wrasses are known to jump from aquariums, so a lid is essential. The eightline is the most aggressive and largest member of its genus. Only one male should be kept, unless the tank is extremely large. Males will also dominate other flasher wrasses if kept together. The best option is to house one male with several female eightline flashers to disperse their aggression. This species will do best with plenty of hiding places in crevices. It should be offered a varied diet of meaty foods, including finely chopped seafoods and mysis shrimp.

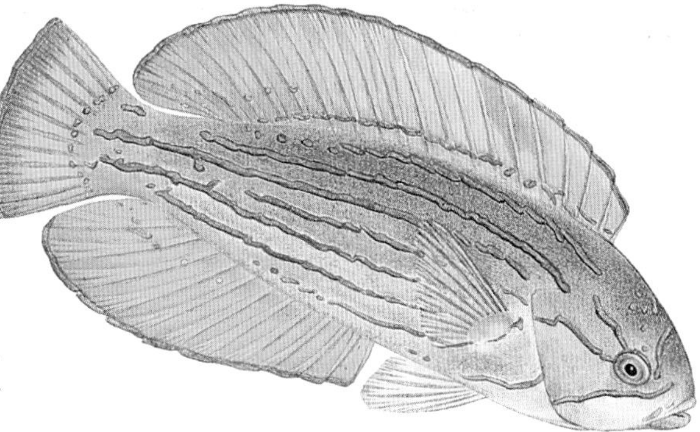

Scientific name	*Paracheilinus octotaenia*
Family	Labridae
Size	5–10cm (2–4in)
Distribution	Red Sea
Habitat	Marine: seaward coral reefs
Diet	Small fish, crustaceans, and shellfish
Breeding	Rarely bred in aquariums

Pearly Jawfish

Also known as the yellow-headed jawfish, this jawfish has a pearly blue body and a golden head. Its long pelvic fins are also golden. Jawfishes frequently jump out of the tank if startled, so a cover is essential, especially at night. The tank should contain plenty of rocks and must have a deep sandy substrate, as this species is a compulsive burrower. The pearly jawfish spends most of its time burrowing its den or hovering vertically outside it, darting back inside when threatened. Rocks in the aquarium should be resting on the bottom of the tank, or this fish could easily cause a rockfall. Pearly jawfish are often kept singly, but several can be kept together as long as they are not crowded. Other tankmates should be docile and similarly sized.

Scientific name	*Opistognathus aurifrons*
Family	Opistognathidae
Size	5–10cm (2–4in)
Distribution	Tropical western Atlantic Ocean
Habitat	Marine: shallow sandy areas on the leeward side of reefs
Diet	Mysis and brine shrimp, plus pellets and flakes
Breeding	100 eggs orally incubated by the male

Comet

The comet, also known as the marine betta, is a deservedly popular fish for the marine reef aquarium: it is hardy while being both beautiful and fascinating to watch. It is brown, with white spots on the head, body, and fins. The spots become smaller and more numerous as the fish ages. An ocellus, or eyespot, is on the back portion of the dorsal fin. When threatened, the comet mimics the aggressive whitemouth moray. It swims into a crevice, leaving its back portion exposed. With its ocelli and white spots, the fish's posterior resembles the face of the moray, scaring off predators. The comet frequently displays this behavior in the home aquarium, swimming backward at the observer. When first introduced to the aquarium, a comet may be very shy and feed poorly until it is established.

Scientific name	*Calloplesiops altivelis*
Family	Plesiopidae
Size	15–20cm (6–8in)
Distribution	Western Pacific Ocean, Red Sea, East African coast, Great Barrier Reef
Habitat	Marine: reefs
Diet	Live foods such as brine shrimp; can be weaned onto prepared frozen foods
Breeding	300–500 eggs deposited on a crevice wall

Emperor Angelfish

Young emperor angelfish are dark blue with electric blue and white rings. After they reach four years old, adults have yellow and blue stripes, with black around the eyes. This change in coloring may not take place in aquariums. This reef species will pick at many corals and damage them, but is a favorite in reef aquariums filled with stony corals with small polyps. Although this species is hardy and adapts well to aquarium life with good care, it is best introduced to a large and established tank with at least six months' growth of green microalgae and plenty of live rock to feed on. As this species can be territorial, particularly with its own species, it will be happiest with plenty of hiding spaces into which it can retreat.

Scientific name	*Pomacanthus imperator*
Family	Pomacanthidae
Size	35–40cm (14–16in)
Distribution	Indian and Pacific oceans: Red Sea to Hawaii
Habitat	Marine: outer coral reefs
Diet	Frozen prepared foods, frozen brine and mysis shrimp, sponges, and dried algae
Breeding	Not bred in aquariums

False Clown Fish

The false clown fish was a popular marine aquarium fish even before it found cinematic fame in the 2003 film *Finding Nemo*. This species is closely related to the true clown fish (*Amphiprion percula*), which is slightly brighter in coloration. The false clown fish is easily recognized by its orange color, three white bars, and black borders on the stripes and fins. In the wild, these anemone-dwelling fish are found in ritteri or malu anemones (*Heteractis magnifica*), brown and purple carpet anemones (*Stichodactyla gigantean*), and green carpet anemones (*Stichodactyla mertensii*), upon which it depends entirely for shelter. The fish are protandrous hermaphrodites, which means that all individuals develop first into males and then possibly into females later. If the single female in a group is removed, the largest male will become a female.

Scientific name	*Amphiprion ocellaris*
Family	Pomacentridae
Size	6–9cm (2.5–3.5in)
Distribution	Eastern Indian and western Pacific oceans
Habitat	Marine: coral reefs, among sea anemones
Diet	Flake food and algae, with live rock growth to graze on
Breeding	Up to 1,000 eggs laid on a sheltered rock

True Percula Clown Fish

Closely related to the false clown fish, this fish can be recognized by its bright orange color and three white bands outlined in black, with black fin markings. The true clown fish has ten spines in its first dorsal fin while the false clown fish has eleven. These fish form lifelong pairs and exhibit courting behavior. Within each fish group dwelling in a host anemone, there will be a breeding pair as well as up to four nonbreeding males. The female is largest, followed by the breeding male, with the size descending down the hierarchy. If the female dies, the largest male changes sex and the remaining males ascend the hierarchy. Like all damselfish, the true clown fish can be territorial and aggressive, especially as it ages. At least a 110-litre (25-gallon) tank will be needed to house it.

Scientific name	*Amphiprion percula*
Family	Pomacentridae
Size	6–9cm (2.5–3.5in)
Distribution	Around northern Australia and Indonesia
Habitat	Marine: coral reefs, among sea anemones
Diet	Flake food and algae, with live rock growth to graze on
Breeding	400–1,500 eggs laid on a sheltered rock

Blacktail Humbug

Also known as the four-striped damsel, the blacktail humbug is one of the easiest marine fish to keep and is ideal for a first-time aquarium owner. The coloration of this damselfish consists of white and black vertical bars covering the length of the fish. There are usually four black bars and three white bars, with a small white spot just above the lip of the fish. Like other damselfish, the humbug can be territorial and feisty, particularly as it ages, so it should be kept with fish of a similar size and watched closely for destructive tendencies. The humbug can live for ten years in a home aquarium if protected from dangerous battles. This species is tolerant of chemical and physical conditions and will eat just about any marine aquarium food offered.

Scientific name	*Dascyllus melanurus*
Family	Pomacentridae
Size	6–8cm (2.5–3in)
Distribution	Eastern Indian and western Pacific oceans
Habitat	Marine: isolated coral heads in sheltered locations
Diet	Algae and live, frozen, and flake foods
Breeding	Up to 600 eggs laid on dead coral or a rock

Sulphur Damsel

This hardy and easy-to-keep reef fish is ideal for first-time marine aquarium owners. The sulphur damsel is bright yellow, with faint vertical bars. The damsels are always popular in reef aquariums because their grazing habits keep coral algae-free without harming it. They also eat the naturally occurring zooplankton, so they should not be kept in an aquarium with excessive mechanical filtration, which will reduce plankton. They will usually accept flakes but will be healthiest when also fed live and freeze-dried food. Although the territorial sulphur damsel will be happy in a community aquarium, it should be kept with fish of a similar size in a spacious aquarium with plenty of cover. Keeping clown fish or butterfly fish in the same tank is not recommended, as these fish will seek out similar hiding places and come into conflict.

Scientific name	*Pomacentrus sulfureus*
Family	Pomacentridae
Size	9–12cm (3.5–4.5in)
Distribution	Western Indian Ocean
Habitat	Marine: coral reefs
Diet	Algae and live, frozen, and flake foods
Breeding	Up to 600 eggs laid on dead coral or a rock

Domestic Goose

Domestic geese may belong to one of two species, the greylag goose (*Anser anser*) or the swan goose (*Anser cygnoides*), or be a hybrid of the two. Geese were first domesticated—for their down feathers, eggs, and meat—at least 7,000 years ago. They have since been bred for their size and fertility. Domestic birds are more than twice as heavy as wild birds, and their relatively fat tail end has given them a more upright posture and prevents them from taking flight. Domestic birds can lay as many as 160 eggs a year, compared to the dozen of the wild goose. Goose eggs are larger than chicken eggs but taste similar. Common domesticated goose breeds include the Toulouse (gray feathers and large dewlaps) and Embden (white feathers and an orange bill).

Scientific name	*Anser anser* or *Anser cygnoides*
Family	Anatidae
Size	Embden 9–14kg (20–30lb); Toulouse 7–9kg (15–20lb)
Distribution	Worldwide
Habitat	Large yard or pasture, with shelter from predators and weather
Diet	Grain, pellets, and grass
Breeding	50–160 eggs a year, depending on breed

Barbary Duck

Often known as the Muscovy duck, the Barbary was first domesticated by various Native American cultures. This breed is popular because its meat is much more lean, tender, and tasty than that of other domesticated ducks, which are descended from the mallard. Barbaries are also heavier than other domesticated ducks. Wild Barbaries have a wide range of feather patterns and colors; white is often the preferred feather color for those raised for meat. All Barbaries have long talons on their feet and a wide tail. The male has a red face with a caruncle, or wattle, at the base of the bill. The ducklings are mostly yellow, but some may have a darker head and blue eyes. The drake has a breathy call, while the hen makes a quiet coo.

Scientific name	*Cairina moschata*
Family	Anatidae
Size	Drakes 4.5–8kg (10–17lb); hens 2.2–4.5kg (5–10lb)
Distribution	Wild populations in the Americas and Europe; domesticated worldwide
Habitat	Wild populations around forested lakes and streams
Diet	Dabbling for plant material in shallow water
Breeding	Clutch of 8–16 eggs up to three times a year

Yellow-Crested Cockatoo

Cockatoos are known for their curved beak, movable head crest, and zygodactyl foot (with two forward-pointing and two backward-pointing toes). This medium-sized white species of cockatoo, also known as the lesser sulfur-crested cockatoo, has a bright yellow crest. The sexes can be easily differentiated: cocks have jet-black eyes while hens have brownish-red eyes. There are twenty-one species in the cockatoo family, which belongs to the wider parrot order. Cockatoos live for thirty to seventy years, so owners must be dedicated to their care. Cockatoos love to cuddle with humans, but should also be encouraged to play alone with toys so that they do not become overly dependent. A large cage and an out-of-cage perch are needed. When highly trained, cockatoos can do tricks and have been used in companion animal therapy.

Scientific name	*Cacatua sulphurea*
Family	Cacatuidae
Size	30–35cm (12–14in) long
Distribution	Indonesia; trade of birds trapped in the wild is illegal
Habitat	Scrub and woodland
Diet	Pellets and fresh fruits and vegetables
Breeding	2–3 eggs in a clutch

Moluccan Cockatoo

The Moluccan, also known as the salmon-crested cockatoo, is one of the most attractive birds in its family. It is pale peachy pink with a crest that it raises when excited or emotional to reveal bright orange-red feathers. This species has a particularly loud call and, like other large cockatoos, can engage in some destructive behavior. It requires a great deal of attention and training to keep it healthy and happy. For these reasons, this species may not be suitable for first-time owners. However, when highly trained, the Moluccan is popular at bird shows. These birds especially enjoy playing with their toys, particularly anything chewable. The Moluccan is an endangered species, so trade in birds trapped in the wild is illegal. Captive-bred birds can be sold with appropriate CITES (Convention on International Trade in Endangered Species) certification.

Scientific name	*Cacatua moluccensis*
Family	Cacatuidae
Size	45–50cm (18–20in) long
Distribution	South Moluccas in eastern Indonesia
Habitat	Lowland forests
Diet	Seeds, nuts, fruit, and coconuts
Breeding	2–3 eggs in a clutch

Red-Tailed Black Cockatoo

This large Australian cockatoo has five regional subspecies, two of which are under threat from habitat destruction, making trade in wild birds restricted by the Australian government. The male has black plumage, a black crest, and two lateral red panels on its tail. Females are brown-black with orange on the tail and chest. Although large cockatoos are not recommended for first-time owners, the red-tailed black cockatoo can be hardy and long-lived if given plenty of space, attention, and training. In the mating season, males will court by puffing up their chest and cheek feathers and hiding their beaks. The male then struts and sings, finishing with a flourish—jumping up to display his red tail feathers. This species has several different calls, from its sharp alarm call to its rolling contact call.

Scientific name	*Calyptorhynchus banksii*
Family	Cacatuidae
Size	58–64cm (23–25in) long
Distribution	Australia
Habitat	Eucalyptus woodlands and around water courses
Diet	Seeds, nuts, berries, and fruits
Breeding	1–2 eggs in a clutch

Galah

This pretty cockatoo is native to Australia. Males and females have a gray back, a strawberry-pink face and chest, and a pale pink crest. The galah has a quieter voice than most cockatoos and can become quite tame, making it one of the easier-to-keep species. Since galahs tend to bond very closely and lovingly with their owners and may well outlive them, adopting a galah as a pet should not be taken lightly. Both male and female galahs are great chatterers, with males just edging ahead with their vocabulary. All galahs like company and clearly see themselves as one of the family, although, as with all cockatoos, overhandling is not recommended. Galahs should be fed a diet low in oil seeds, as they are susceptible to fatty tumors.

Scientific name	*Eolophus roseicapillus*
Family	Cacatuidae
Size	33–36cm (13–14in) long
Distribution	Australia
Habitat	Open habitats to urban areas
Diet	Seeds (with few oil seeds), nuts, berries, and fruits
Breeding	2–5 eggs in a clutch

Cockatiel

The smallest birds in the cockatoo family, cockatiels are the second most popular caged bird, after the budgerigar. Males and females usually have gray plumage with orange patches on the ear areas. The face of the male is yellow or white, while the face of the female is gray. However, there are many different naturally occurring color mutations. In captivity, a cockatiel will commonly live for fifteen to twenty years. Well-socialized birds are usually sweet-natured and friendly. They may even enjoy taking a nap on their owner. A cockatiel that would like its head to be scratched will lower it expectantly—and then give a squawk to express its approval. Cockatiels are better at mimicking whistles and sounds than speech. They may learn to mimic household noises such as ringing phones, washing machines, and flushing toilets.

Scientific name	*Nymphicus hollandicus*
Family	Cacatuidae
Size	30–33cm (12–13in) long
Distribution	Inland Australia
Habitat	Bush and scrub close to water
Diet	Seeds, nuts, berries, and fruits
Breeding	4–5 eggs in a clutch

Domestic Pigeon

The domestic pigeon, a descendant of the rock pigeon, was domesticated at least 5,000 years ago. Pigeon fanciers, as pigeon owners are known, may keep their birds for homing, exhibiting in shows, or taking part in flying and sporting competitions. There are more than 300 breeds of "fancy" pigeons, developed for particular attributes. Birds bred for their ornamental feathers include the fantail, frillback, and Oriental frill. A breed with an unusual laughing voice is the trumpeter. Homing pigeons include the dragoon and English carrier. Those with acrobatic and sporting abilities include the Danzig highflyer and the roller. Pigeons must live in specially designed lofts that allow the birds to come and go as they choose and which contain perches, nesting boxes, and separate pens as needed. An aviary will be necessary for birds that are not at complete liberty to fly.

Scientific name	*Columba livia*
Family	Columbidae
Size	28–36cm (11–14in) long
Distribution	Worldwide
Habitat	All regions and climates
Diet	Grains and seeds, plus some greens
Breeding	2 eggs in a clutch

Orange-Cheeked Waxbill Finch

This sprightly finch has a prominent orange marking on the sides of its face, crimson rump feathers, and an orange-red beak. Females look similar to males, with paler orange cheek patches. These birds need a large "flight," or aviary, with shrubbery and branches for roosting alongside tall grasses and reeds. In addition to a suitable seed mix, this finch should be offered small mealworms, fruit flies, and pinhead crickets every day. In an enclosure shared with other, more aggressive species, these sweet-natured birds may be intimidated if only one feeding station is provided. These finches commonly make high-pitched peeps, while the male has a pleasant, low song. The orange-cheeked waxbill is an acrobatic and entertaining bird to watch, often hanging upside down while feeding and jumping up vertical branches.

Scientific name	*Estrilda melpoda*
Family	Estrildidae
Size	8–10cm (3–4in) long
Distribution	West Africa
Habitat	Grassland and scrub
Diet	Seed mixes and insects
Breeding	3–6 eggs in a clutch

Blue-Capped Cordon-Bleu Finch

There are five species of cordon-bleu finch: the blue-capped, blue-breasted, and red cheeked cordon-bleus, and the purple and common grenadiers. All of the species are native to Africa and are seed eaters with short, pointed bills. In the wild, they build large, domed nests. These finches are sensitive to the cold and are difficult to establish in the home, but can live up to fifteen years in the right conditions. They will do best in a large planted aviary containing mixed species. Cordon-bleus are difficult to breed in the home because they need plenty of privacy. The cordon-bleus also require a constant supply of insects to eat. Fruit fly larvae, ant eggs, mealworms, and waxworms should be offered daily. Blue-capped cordon-bleus are known for their lovely call, and both sexes will call and dance.

Scientific name	*Uraeginthus cyanocephalus*
Family	Estrildidae
Size	10–13cm (4–5in) long
Distribution	Tanzania, Sudan, Somalia, Kenya, and Ethiopia
Habitat	Grassland and desert
Diet	Seed mixes, plenty of small insects, and some greens
Breeding	4–6 eggs in a clutch

Canary

First bred in captivity more than 300 years ago, the pet canary is a domesticated form of the wild bird, a member of the finch family, which originated in the Canary Islands, the Azores, and Madeira. There are many different breeds of canaries, developed for their color, such as the bronze, ivory, and red factor; for their shape, such as the Fife, Gibber Italicus, and Raza Española; or for their unique song patterns, such as the Spanish timbrado, American singer, or Persian singer. The oldest canary breed is the lizard canary, which was first recognized in 1709. Its speckled plumage resembles the scales of a lizard. Canaries are popular pets because of their singing ability. Some birds take part in singing contests where recognized "tours" and "rolls" must be performed.

Scientific name	*Serinus canaria*
Family	Fringillidae
Size	10–18cm (4–7in) long
Distribution	Wild birds in Canary Islands, Azores, and Madeira
Habitat	Wide range of habitats, from sand dunes to gardens
Diet	Seeds mixes, some greens, and small insects
Breeding	3–5 eggs in a clutch

Domestic Turkey

The modern domesticated turkey is a descendant of the wild turkey (*Meleagris gallopavo*). The average natural life span of a domesticated turkey is ten years. Most have white feathers, but various brown and bronze varieties are also kept. The broad-breasted white turkey is the most common choice on turkey farms. The turkey that receives the annual traditional pardon from the U.S. president belongs to this breed. All turkeys have a fleshy protuberance called a wattle on the underside of the beak. Although turkeys are normally kept for their tasty and healthy meat, there is a long tradition of turkeys as companion animals in the United States. Abraham Lincoln's son Tad kept a turkey named Jack in the White House. Many owners claim they are bright and sociable animals.

Scientific name	*Meleagris gallopavo*
Family	Meleagrididae
Size	Males 8kg (18lb); females 3.5kg (8lb)
Distribution	Native to North America; kept domestically worldwide
Habitat	Domesticated; wild turkeys live in woods and fields
Diet	Grasses, nuts, seeds, roots, and insects
Breeding	10–14 eggs in a clutch

Ko-Shamo Chicken

With a population of more than 24 billion, the chicken is the most popular domestic animal in the world. The shamo is an attractive Japanese chicken and is a popular show bird all over the world. *Shamo* means "fighter" in Japanese, and the chicken was bred for this. There are several breeds of shamos, with the ko-shamo being a small variety. All shamos can be very tame if treated considerately, even begging and following for food. Chickens are generally easy and inexpensive to care for. They are usually kept in a roost at night and in a pen during the day, unless they are free-range. They must always be protected from predators. Roosters can become aggressive and need their own pen—but hens will still lay eggs without the presence of a rooster, although such eggs are unfertilized and will not hatch.

Scientific name	*Gallus gallus*
Family	Phasianidae
Size	0.8–1kg (1.7–2.2lb)
Distribution	Originated in Japan; check urban regulations on keeping chickens and roosters
Habitat	Domesticated
Diet	Balanced commercial feed plus plants, seeds, insects, and fruit and vegetable peelings
Breeding	Around 12 eggs in a clutch

Dorking Chicken

The Dorking originated in Italy at least 2,000 years ago. This versatile breed is kept both for egg and meat production. It was developed for its good-tasting meat. The Dorking has a rectangular body and short legs with five toes. Dorkings have a large comb and produce a white-shelled egg. Varieties include white, silver-gray, colored, and the rare cuckoo rosecomb. All varieties may need protection in cold weather.
A chicken usually lays an egg a day until it has a clutch of around twelve eggs. It then does not lay for a couple of days before beginning another clutch. Under natural conditions, hens lay a clutch, become "broody," and then incubate the eggs until the chicks hatch. Dorking hens are docile and will frequently become broody; they make good mothers.

Scientific name	*Gallus gallus*
Family	Phasianidae
Size	3–4kg (7–9lb)
Distribution	Originated in Italy; check urban regulations on keeping chickens and roosters
Habitat	Domesticated
Diet	Balanced commercial feed plus plants, seeds, insects, and fruit and vegetable peelings
Breeding	Around 12 eggs in a clutch

Fischer's Lovebird

This small parrot, named after the nineteenth-century German explorer Gustav Fischer, is well known for its gorgeous multicolored plumage. It has a green back, chest, and wings; a golden neck that progresses to orange along the face; and an olive top to the head. These highly sociable birds should always be kept with a companion in a roomy cage or aviary. Lovebirds are very active and will enjoy plenty of toys, objects to chew on, and a birdbath for frequent dips. Once a lovebird is settled in the home, it will become curious and will enjoy getting out of its cage to fly around, provided that it is protected from hazards. Although Fischer's lovebirds do not like to be touched by humans, they may happily perch on their owners if given a snack.

Scientific name	*Agapornis fischeri*
Family	Psittacidae
Size	14cm (5.5in) long
Distribution	Tanzania
Habitat	Trees on grassy plains
Diet	Seeds, vegetables, fruit, and grains
Breeding	3–8 eggs in a clutch

Turquoise-Fronted Amazon Parrot

This talkative bird—which may live to be nearly a hundred years old—is commonly kept as a pet around the world. Both an excellent mimic and singer, the turquoise-fronted Amazon, also known as the blue-fronted Amazon, is far from shy and is happy to perform for family and strangers alike. It is a playful, interactive bird, although it is also content to entertain itself alone with its toys. Turquoise-fronted Amazons need plenty of room and suitable perches. This green-feathered bird has an eponymous blue marking on its head just above its beak, and yellow on its face and crown. The shoulder edges of the wings are red. Unlike other species of Amazon parrots, the turquoise-fronted has been less affected by deforestation in its habitat and is not yet at risk.

Scientific name	*Amazona aestiva*
Family	Psittacidae
Size	38cm (15in) long
Distribution	Bolivia, Brazil, Paraguay, and northern Argentina
Habitat	Forests and palm groves
Diet	Seed mixes, fruit, and some greens
Breeding	3–5 eggs in a clutch

Blue-and-Yellow Macaw

This striking bird is among the largest parrots. Macaws are Central and South American parrots that are distinguished by their large, dark beaks and relatively hairless, light-colored facial patches. The powerful beaks and harsh, loud voices of many larger macaws tend to make them unsuitable for most first-time bird owners. For someone who can meet its needs, however, the blue-and-yellow macaw can make an excellent and loving companion bird. With proper training, this macaw can develop an extensive vocabulary. The blue-and-yellow macaw has a bright blue tail and wings, a dark blue chin, yellow underparts, and a green forehead. Its naked white face, which turns pink when excited, is surrounded by small black feathers. Its black beak is used for crushing nuts and hanging from trees in the wild.

Scientific name	*Ara ararauna*
Family	Psittacidae
Size	76–86cm (30–34in) long
Distribution	South America: Trinidad and Venezuela to Brazil and Paraguay Central America: Panama
Habitat	Bush and scrub close to water
Diet	Pellets, seeds, nuts, flowers, and fruit
Breeding	2–3 eggs in a clutch

Scarlet Macaw

The scarlet macaw typically lives for thirty to fifty years, although some birds can survive for up to seventy-five years. This species' plumage is largely scarlet with blue rump and tail feathers, and yellow and blue on the wings. Birds display bare white skin from around the eyes to the bill. Scarlet macaws are an expensive and high-maintenance pet, but will reward careful and attentive owners with love and intelligence. These birds require extremely large cages with room to spread their wings, and plenty of out-of-cage time. Frequent interaction with their owners is essential for them to remain tame and friendly. They love to play and need to chew to protect against beak overgrowth, so stimulating and chewable toys are a necessity. Macaws enjoy water and may have fun taking a bath in the kitchen sink.

Scientific name	*Ara macao*
Family	Psittacidae
Size	80–90cm (32–36in) long
Distribution	Central America: Mexico, Panama, Guatemala, and Belize South America: Amazon Basin to Paraguay
Habitat	Rain forest, woodland, savanna, and around water courses
Diet	Pellets, seeds, nuts, vegetables, and fruit
Breeding	2–3 eggs in a clutch

Military Macaw

This medium-sized macaw has mostly green plumage with blue flight feathers on its wings and a red tail bordered with blue. It has a red patch above its nostrils and a naked white facial area barred with black. As with all parrots, it is essential to ensure that your military macaw was bred in captivity, because trade in wild birds is illegal. With training and devotion, the military macaw can become a rewarding pet that may live for as long as sixty years. These birds enjoy human interaction and can develop a large vocabulary. A military macaw needs a cage that is a minimum of 120 by 90 centimetres (48 by 36 inches) in area and 150 centimetres (60 inches) high. Plenty of out-of-cage time is also essential. Since macaws like to stretch their feet, a variety of perch sizes will be enjoyed.

Scientific name	*Ara militaris*
Family	Psittacidae
Size	68–78cm (27–31in) long
Distribution	Mexico to Argentina
Habitat	Forest and woodland
Diet	Pellets, seeds, nuts, vegetables, and fruit
Breeding	1–2 eggs in a clutch

Burrowing Parakeet

The burrowing parakeet, also known as the Patagonian conure, is the largest member of the South American conure group. The conures are distinguished by their breasts, which appear scaly. They are often known as the clowns of the parrot world due to their attention-seeking antics, including hanging upside down and appearing to dance by swaying back and forth. The burrowing parakeet is highly attractive but can be noisy and destructive with its chewing. It is largely gray with stunning orange and red underparts. Well-kept birds are known for being sweet-natured, affectionate, and intelligent. They have a clear, high-pitched voice and are capable of talking. Burrowing parakeets get their name from the fact that in the wild they sleep and breed in burrows dug into sandstone cliffs. This species pairs for life, and both parents care for the young.

Scientific name	*Cyanoliseus patagonus*
Family	Psittacidae
Size	46–53cm (18–21in) long
Distribution	Argentina, Chile, and Uruguay
Habitat	Arid bush steppe
Diet	Pellets, seeds, fruit, grains, green vegetables, and cuttlefish bones
Breeding	3–8 eggs in a clutch

Eclectus Parrot

Males and females of this species are widely different in appearance, exhibiting the most extreme sexual dimorphism of all psittacines. The gorgeous females have red heads and blue to purple breasts, with black beaks. The males are bright green with blue or red tail and wing feathers, and sport orange upper mandibles with yellow tips and black lower mandibles. When they were first observed, it was thought that the cocks were a different species from the hens. Cocks tend to make better pets than the more aggressive hens. Neither sex will become a proficient talker. These parrots can settle well in the home, provided they have plenty of opportunities to fly, preferably in a flight cage. Ten *Eclectus* subspecies have been observed, such as the Solomon Island and the red-sided, with differences in coloration, size, and habitat.

Scientific name	*Eclectus roratus*
Family	Psittacidae
Size	33–36cm (13–14in) long
Distribution	Northeastern Australia, New Guinea, and Solomon and Maluku islands
Habitat	Rain forest canopy
Diet	Seeds, nuts, vegetables, greens, and fruit
Breeding	2–4 eggs in a clutch

Black-Capped Lory Parrot

These beautifully colorful birds, which belong to the true parrot family, need a little more care than the average parrot due to their diet, which consists primarily of nectar supplemented with fruit and a little dried seed. This liquid diet causes relatively fluid droppings, which can be a drawback around the home. Like lorikeets, lories have brush-tipped tongues specialized for feeding on nectar. There are seven subspecies of black-capped lory, all of which have a red head with a black cap, green wings with yellow underwings, and blue legs and belly. Youngsters have horn-colored bills, which turn orange as they grow. This species of lory has a particularly loud call and can be a highly successful mimic. In the wild, the black-capped is usually found in pairs or groups, so it will be happiest with a companion bird.

Scientific name	*Lorius lory*
Family	Psittacidae
Size	30cm (12in) long
Distribution	Java to New Guinea
Habitat	Rain forest
Diet	Nectar, fruit, and dried seeds
Breeding	4–6 eggs in a clutch

Budgerigar

Commonly also called a parakeet, the budgerigar is a small, broad-tailed parrot. It is probably the most common caged bird in the world. The most frequently seen colors for "budgies" in captivity are blue, green, and yellow. These intelligent and sociable birds enjoy interacting with each other, with humans, and with toys. Pets can be taught to speak and whistle tunes, with males being significantly better at these skills than females. Females can usually mimic no more than a dozen words, while highly trained males can master up to a hundred. Caged budgies usually live for five to eight years but have been known to survive for twenty years if very well cared for. An ideal cage size for two birds would be at least 75 centimetres (30 inches) long, and plenty of out-of-cage exercise should be offered.

Scientific name	*Melopsittacus undulatus*
Family	Psittacidae
Size	18cm (7in) long
Distribution	Australia
Habitat	Scrub and grassland
Diet	Seeds, grains, edible flowers, vegetables, and fruit
Breeding	4–8 eggs in a clutch

Crimson Rosella

This Australian parrot is predominantly red, with blue cheeks and tail, and black-scalloped, blue-edged wings. Juveniles have greenish plumage that "ripens" to red. In the wild, crimson rosellas are nonmigratory. Outside the breeding season, large groups of birds will chatter together as they forage for food. During the breeding season, these birds—which are monogamous—will forage only with their mate. Rosellas love to fly, so in captivity they will do best in a single-species aviary or long flight cage. Single pets can be housed in a medium-sized parrot cage as long as they are allowed plenty of out-of-cage time. Rosellas should be given water to bathe in, as this is something that they frequently enjoy. This species does not have much talking ability, but birds can mimic whistles and songs.

Scientific name	*Platycercus elegans*
Family	Psittacidae
Size	34–37cm (13.5–14.5in) long
Distribution	Eastern and southeastern Australia; introduced to New Zealand
Habitat	Forests and woodland
Diet	Seeds, nectar, berries, fruit, and nuts
Breeding	3–8 eggs in a clutch

Red-Rumped Parrot

Well known for its pretty plumage, the red-rumped parrot—known variously as the crimson-winged parrot, red-backed parrot, and red-winged parrot—is native to Australia. The male's gorgeous plumage is emerald green with yellow underparts, and blue on the wings and upper back. Only the male has the red rump. The female is less vibrant, with green wings and back, olive underparts, and black wing tips. Red-rumped parrots are not renowned for their talking ability, but they can be encouraged to do so with patience and plenty of training. This species needs daily handling to become socialized and remain tame. Males should not be housed together, and red-rumped parrots are not tolerant of other species. These birds enjoy cage toys and taking baths—which are a perfect opportunity to splash everyone in reach.

Scientific name	*Psephotus haematonotus*
Family	Psittacidae
Size	27–29cm (10.5–11.5in) long
Distribution	Southeastern Australia
Habitat	Open land close to water
Diet	Seeds, nuts, vegetables, and fruit
Breeding	4–6 eggs in a clutch

Plum-Headed Parakeet

Males of this species, part of the genus *Psittacula*, have a plum-colored head with a narrow black collar around their neck. They have dark green wings, while the rest of the body is light green. They display a red shoulder patch and a blue-green rump and tail, tipped with white. The female's head is gray and she lacks the collar and shoulder patch. These are highly sociable birds and will tolerate being in an aviary of mixed species unless they are breeding, when they can become more aggressive. These birds should have plenty of time and space to fly as well as a wide variety of chewable toys. This species has a variety of melodic calls, a shrill note when in flight, and a potentially loud scream.

Scientific name	*Psittacula cyanocephala*
Family	Psittacidae
Size	32–34cm (12.5–13.5in) long
Distribution	Sri Lanka, India, Bangladesh, and Pakistan
Habitat	Forest and woodland
Diet	Seeds, nuts, blossoms, leaf buds, and fruit
Breeding	4–6 eggs in a clutch

Alexandrine Parakeet

This species is named after Alexander the Great, who is said to have been one of the first people to export parrots from India to Europe. The male Alexandrine parakeet's most noticeable feature is its black neck ring and the pink band around the nape of its neck. It has a red bill tipped with yellow and a green head with grayish-blue cheeks. It displays a dark red shoulder patch and a blue-green tail tipped with yellow. Females lack the pink-and-black collar and are duller in color. In the wild, Alexandrine parakeets are sociable birds, forming huge, noisy flocks at dusk. These birds are relatively expensive to buy, but hand-reared birds are popular for their tameness and ability to talk. They need a wide and tall cage and plenty of chewable toys to keep them stimulated.

Scientific name	*Psittacula eupatria*
Family	Psittacidae
Size	56–60cm (22–24in) long
Distribution	Bangladesh, India, and Sri Lanka to Southeast Asia
Habitat	Forest, woodland, and fields
Diet	Seeds, nuts, blossoms, and fruit
Breeding	2–4 eggs in a clutch

Rose-Ringed Parakeet

The rose-ringed parakeet, also known as the ringnecked parakeet, was first brought to Europe by the ancient Greeks, and has remained a popular pet ever since. Adult males have the characteristic black neck ring and pink band around the nape of the neck. Females and juvenile birds either have no neck rings or display pale gray rings and faded nape bands. The Indian subspecies of the rose-ringed parakeet has established large feral populations of escaped and introduced birds in many European cities, Japan, Florida, and Israel. A particularly large population center of such birds exists in southeastern England, where it is believed that the numbers of parakeets could endanger native British birds. Rose-ringed parakeets are usually hardy pets and need a little less human interaction than other parakeets to keep them well adjusted and tame.

Scientific name	*Psittacula krameri*
Family	Psittacidae
Size	38–42cm (15–16.5in) long
Distribution	African subspecies: Guinea to Sudan Indian subspecies: India, Pakistan, and Nepal
Habitat	Forests to fields
Diet	Seeds, nuts, fruit, vegetables, and buds
Breeding	2–4 eggs in a clutch

African Gray Parrot

The African gray is considered to be the best talker of all the parrots and, as a result, has been a popular pet for centuries. These birds are gray apart from the tail, which is bright red. This is a highly intelligent species, probably the brightest of all birds. Some owners believe that their birds can associate the many words they mimic with their meanings and respond to their owners with a humanlike intelligence. To engage their lively brains, birds should be provided with puzzle toys and foraging toys, in which they must learn how to extract food from the toy. These parrots need plenty of attention as well as several hours of out-of-cage time a day. Although they do not get along well with other species, birds benefit from being kept with other African grays.

Scientific name	*Psittacus erithacus*
Family	Psittacidae
Size	33–40cm (13–16in) long
Distribution	Western and central Africa
Habitat	Rain forest
Diet	Seeds, nuts, vegetables, and fruit
Breeding	3–5 eggs in a clutch

Pesquet's Parrot

This endangered parrot is native to New Guinea. As with all parrots, trade in wild birds is illegal. The Pesquet's parrot is black with a red belly. It has bare black facial skin and a fairly long, hooked bill, giving the bird a vulturelike profile—and providing it with its alternative name, the vulturine parrot. The bare face is an adaptation to avoid feather matting from the sticky figs that are this bird's sole food. In the wild, these parrots are seasonally nomadic as they search for figs. As with other parrots, care should be taken that a pet does not develop the habit of plucking its feathers. Precautions should include not leaving the bird alone for long periods, and making sure that it has an adequate diet.

Scientific name	*Psittrichas fulgidus*
Family	Psittacidae
Size	43–48cm (17–19in) long
Distribution	New Guinea
Habitat	Forest
Diet	Figs
Breeding	2 eggs in a clutch

Thick-Billed Parrot

A bright green bird, the thick-billed parrot has a large black bill and a red forehead, shoulders, and thighs. It makes a range of loud, rolling calls. In the wild, the calls of a flock resemble human laughter and can be heard from a long way away. In its natural habitat, the thick-billed parrot feeds almost exclusively on pine seed, and the bird's life is centered on pinecone production. Birds are nomadic, following areas of cone abundance, and breed at times of peak production. The thick-billed parrot is endangered due to logging and illegal capture for the pet trade. It is one of only two species of parrot once native to the United States; the other, the Carolina parakeet, is now extinct. Luckily, there are numerous captive breeding programs. In captivity, these friendly parrots make peaceful and amicable aviary birds.

Scientific name	*Rhynchopsitta pachyrhyncha*
Family	Psittacidae
Size	35–40cm (14–16in) long
Distribution	Sierra Madre Occidental region of Mexico
Habitat	Mountain conifer forests
Diet	Pine seeds, acorns, and some buds
Breeding	2–3 eggs in a clutch

Rainbow Lorikeet

The plumage of these gorgeously vivid birds features every color of the rainbow. There are more than twenty subspecies, each of which sports a slightly different coloration. The best known subspecies, *Trichoglossus haematodus moluccanus*, or Swainson's lorikeet, has a dark blue or violet head and stomach, a bright green back and tail, and an orange breast and beak. Overall, the rainbow lorikeet is considered common, but the populations of some subspecies are threatened by habitat loss and the wild-bird trade, so status and provenance should be checked before purchase. Lorikeets make entertaining and affectionate pets, but their relatively demanding nature and life span of more than twenty years mean that ownership should not be taken lightly. Lorikeets have particularly liquid droppings, which need to be swiftly cleaned from furnishings, carpets, and curtains.

Scientific name	*Trichoglossus haematodus*
Family	Psittacidae
Size	25–30cm (10–12in) long
Distribution	Indonesia to southeastern Australia
Habitat	Forest, woodland, and gardens
Diet	Custom mixtures of cereal and pollen, honey and water, and fruit
Breeding	2–3 eggs in a clutch

Toco Toucan

The toco toucan is the largest and best known species in the toucan family. Its orange bill is 20 centimetres (8 inches) long with a black spot near the tip. Its plumage is black with a white throat, edged beneath with red. This species cannot be housed effectively in the average home because it must have plenty of flying space, so a large aviary is essential. Toucans can be aggressive with birds of other species in their family and with smaller species. Toucans cannot mimic human speech, but are playful and inquisitive. They may enjoy sitting on their owner's shoulder or even nestling in a lap. Highly trained toucans can be taught quite complex tricks, including playing catch. When it sleeps, a toucan rests its bill on its back, folding its tail neatly over it.

Scientific name	*Ramphastos toco*
Family	Ramphastidae
Size	56–66cm (22–26in) long
Distribution	Guiana to Bolivia and northern Argentina
Habitat	Semiopen habitats, from woodland to fields
Diet	Pellets, fruits, mealworms, and a small daily ration of minced beef
Breeding	2–4 eggs in a clutch

Sheltie Guinea Pig

Since its domestication about 7,000 years ago, many breeds of guinea pigs have been developed. The sheltie is a long-haired breed. The hair flows back from its face, unlike that of the Peruvian guinea pig, which is the progenitor of all the long-haired breeds and often has its topknot tied back so it can see. The sheltie's hair may grow to as long as 50 centimetres (20 inches), but is usually kept trimmed in animals that are not for show. There is a wide range of colors and patterns, including agouti (in which the root and tip of the hair are different colors) and tortoiseshell (patches of red and black). Like all guinea pigs, the sheltie prefers to be kept in a group of two or more. Groups of females, or females with a neutered male, are good combinations.

Scientific name	*Cavia porcellus*
Family	Caviidae
Size	0.7–0.9kg (1.5–2 lb); 20–25cm (8–10in) long
Distribution	Domesticated in South America; kept worldwide
Habitat	Grassy plains in the wild
Diet	Fresh grass hay, food pellets, raw vegetables, and fruits
Breeding	Average of 3 pups up to five times a year

Smooth-Haired Guinea Pig

Despite its name, the guinea pig is neither a pig nor is it from Guinea. It is a species of rodent that was first domesticated for food by the Andean peoples of South America. The "guinea" portion of the name may be from the fact that, when the animals were brought to Europe in the sixteenth century, "guinea" was often used to describe any distant place. Smooth-haired guinea pigs have fur and a body structure most closely resembling wild ancestors. A wide range of single colors and patterns are available, from tricolored to the spotted Dalmatian.

Guinea pig activity is scattered evenly through the day and night. Animals make a complex array of sounds in response to each other and their owners, including squealing when they are distressed and purring when happy.

Scientific name	*Cavia porcellus*
Family	Caviidae
Size	0.7–0.9kg (1.5–2 lb); 20–25cm (8–10in) long
Distribution	Domesticated in South America; kept worldwide
Habitat	Grassy plains in the wild
Diet	Fresh grass hay, food pellets, raw vegetables, and fruits
Breeding	Average of 3 pups up to five times a year

Abyssinian Guinea Pig

Abyssinian guinea pigs, also known as rosettes, have longer and coarser fur than the smooth-haired guinea pig. This fur forms "rosettes" all over their body, with raised ridges occurring wherever rosettes meet. An ideal show Abyssinian should have a symmetrical pattern of ten rosettes, with one on each shoulder, four across the back, one on each hip, and two on the rump—but much more leeway is given to ordinary pets. As with all guinea pigs, it is worth remembering that the Abyssinian is shy and will try to escape whenever it is picked up. The best way to hold a guinea pig is to grasp it behind the front legs, enclosing its back in your palm. Your other hand should support its stomach. For extra safety, cradle the guinea pig to your chest.

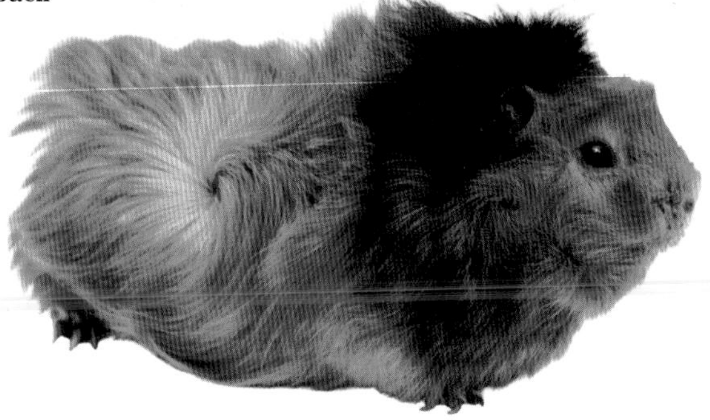

Scientific name	*Cavia porcellus*
Family	Caviidae
Size	0.7–0.9kg (1.5–2 lb); 20–25cm (8–10in) long
Distribution	Domesticated in South America; kept worldwide
Habitat	Grassy plains in the wild
Diet	Fresh grass hay, food pellets, raw vegetables, and fruits
Breeding	Average of 3 pups up to five times a year

Long-Tailed Chinchilla

The captive-bred long-tailed chinchilla is a popular pet, but in its native Chile its population is declining due to hunting for its fur. Trade in wild chinchillas is illegal. The long-tailed chinchilla has large ears, black eyes, and a bushy tail. It is covered in very thick, soft fur to keep it warm in the mountainous regions that are its natural habitat. The chinchilla is an agile runner and jumper. It is primarily a nocturnal animal with its periods of peak activity being dusk and dawn. Pet chinchillas are relatively easy to look after, rarely aggressive, and very clean. Once they are comfortable with their owner, they can be very affectionate and love to be scratched under the chin. Domesticated chinchillas may live up to twenty years, so adopting one as a pet should not be undertaken lightly.

Scientific name	*Chinchilla lanigera*
Family	Chinchillidae
Size	0.3–0.6kg (0.75–1.25lb); 22–38cm (8.5–15in) long, excluding tail
Distribution	Northern Chile
Habitat	Barren mountain slopes
Diet	Chinchilla pellets and fresh hay
Breeding	2–3 young twice a year

Dzungarian Hamster

Dzungarian hamsters are dwarf hamsters, usually of the Campbell's Russian dwarf hamster species, but the term is also used to describe the closely related winter white Russian dwarf hamster. Dwarf hamsters are popular with pet owners who are looking for a smaller, more sociable hamster. While Campbell's hamsters come in a variety of colors and markings, the winter white is gray with a darker dorsal stripe, fading to white in the winter to hide it from predators on the snow-covered steppes of its central Asian habitat. Dzungarian hamsters have a life span of eighteen to thirty-six months and live happily in captivity as long as they are not excessively held—for this reason, they are inadvisable pets for very young children. Although Dzungarians are sociable and can be kept in colonies, mixed groups are best avoided because they are lively breeders.

Scientific name	*Phodopus campbelli* and *Phodopus sungorus*
Family	Cricetidae
Size	7–10cm (3–4in) long
Distribution	Central and eastern Asia
Habitat	Asian steppes
Diet	Suitable seed mixes, food pellets, and fresh vegetables
Breeding	4–6 pups several times a year

Golden Hamster

Also known as the Syrian hamster, the golden hamster is a popular household pet. In the wild, the golden hamster is officially endangered, but domestic pets are bred for the purpose and do not affect the wild population. Like most hamsters, the golden hamster stores food in its extendable cheek pouches. In the natural habitat, this food is transported to the burrow, where as much as 25 kilogrammes (55 pounds) can be stored. Originally, these hamsters came in just brown, black, or gold, but today pet stores are full of golden hamsters in a wide range of shades. Like all hamsters, these are crepuscular animals, which means they are most active at dusk and dawn. Children may enjoy watching a hamster's activities, which include wheel running, scent marking, and grooming. These hamsters are extremely territorial and should be kept individually.

Scientific name	*Mesocricetus auratus*
Family	Cricetidae
Size	13–25cm (5–7in) long
Distribution	Syria
Habitat	Desert borders, scrub
Diet	Suitable seed mixes, food pellets, and fresh vegetables
Breeding	Average of 7 pups (but up to 24) several times a year

Alaska Rabbit

Despite its name, the Alaska rabbit originated in Germany in the 1920s. Bred originally for its fur, this rabbit's silky, jet-black coat is its most notable feature. The hair's undercolor is slate blue with the last centimetre of the hair tipped black. The rabbits sometimes have white hairs interspersed through their coats. To prevent the hair from becoming matted and uncomfortable, this rabbit should be groomed regularly. The Alaskan's body is well muscled and stocky, with almost no neck and the head carried very close. The ears, which are held straight and open, are brown-black. The Alaskan's eyes are dark brown. Owners describe Alaskans as having strong personalities, and are kept busy by their entertaining antics. The Alaskan is a popular pet and regularly appears at shows.

Scientific name	*Oryctolagus cuniculus*
Family	Leporidae
Size	3–4kg (6.5–9lb)
Distribution	Originated in Germany
Habitat	Domesticated
Diet	Grasses, leaves, vegetables, and seeds
Breeding	5–6 babies

Giant Angora Rabbit

As can be seen at first glance, the Angora rabbit was bred for its long, soft fur. It is one of the oldest types of domestic rabbits, originating in Ankara, Turkey, along with the Angora goat and Angora cat. Frequent grooming is necessary to prevent the fur from matting, which causes the animal discomfort. For this reason, Angoras are not recommended for inexperienced rabbit owners. Some Angoras are shorn in summer so that they do not overheat. There are many different Angora breeds, with the five most common being the English, French, giant, satin, and German. The giant Angora produces more wool than the others and is often farmed. Although Angoras require careful handling, most pets are calm and docile.

Scientific name	*Oryctolagus cuniculus*
Family	Leporidae
Size	3.8–5.4kg (8.5–12lb)
Distribution	Originated in central Turkey
Habitat	Domesticated
Diet	Grasses, leaves, vegetables, and seeds
Breeding	3–6 babies

Checkered Giant Rabbit

The checkered giant is a large and active spotted rabbit. Males tend to be less territorial than females and make better pets. The breed originated in Germany as a cross between the Flemish giant and the German checker. It was imported into the United States in 1910, where it has since developed into a racier type distinct from European checkered giants. The checkered giant has distinctive markings. It must have a butterfly-shaped spot on its nose, circles around its eyes, spots on its cheeks, dark ears, a stripe on its spine, and spots on its sides. The markings may be either blue or black, and the rest of the body is white. Because the checkered giant is larger than average, it needs a bigger cage and more feed than some of the smaller breeds.

Scientific name	*Oryctolagus cuniculus*
Family	Leporidae
Size	4.5–6.8kg (10–15lb)
Distribution	Originated in Germany
Habitat	Domesticated
Diet	Grasses, leaves, vegetables, and seeds
Breeding	11–14 babies

Dutch Rabbit

Despite its name, the ever-popular Dutch rabbit originated in England. In the nineteenth century, consignments of live rabbits were regularly shipped from the Netherlands to the meat markets in London. Among these rabbits was the dwarf Brabancon, hailing from Brabant in Flanders. In the 1890s, a Londoner by the name of Copeman, founder of the Dutch Rabbit Club, began picking out dwarf Brabancons with the particular white markings that appealed to him. It was these rabbits that formed the foundation of the breed. The Dutch displays a white blaze on its face, starting on the forehead and encircling the nose, throat, and neck. Six colors are accepted in conjunction with white: black, blue, chocolate, brown-gray, steel gray, and tortoiseshell. Dutch rabbits have a compact, rounded body; spoon-shaped ears; and short, glossy fur.

Scientific name	*Oryctolagus cuniculus*
Family	Leporidae
Size	1.6–2.5kg (3.5–5.5lb)
Distribution	Originated in the United Kingdom
Habitat	Domesticated
Diet	Grasses, leaves, vegetables, and seeds
Breeding	3–5 babies

Dwarf Hotot Rabbit

With its docile, affectionate, and playful nature, this small rabbit is reported to make an excellent children's pet. The dwarf hotot was developed almost simultaneously by two different German breeders in the 1970s. It is a pure white rabbit with highly distinctive eye bands forming a thick outline around each eye, giving the appearance of spectacles. Eye bands may be in black or chocolate, while the rabbit's bright, round eyes are dark brown. This rabbit is compact and stocky, with a round head, short ears, and no visible neck. The dwarf hotot's fur is soft and lustrous. The hotot, which was originally developed in France, is also available in a standard size, known as the blanc de hotot or pharaoh rabbit, and in a rexed, or curly haired, variety.

Scientific name	*Oryctolagus cuniculus*
Family	Leporidae
Size	0.9–1.4kg (2–3lb)
Distribution	Originated in Germany
Habitat	Domesticated
Diet	Grasses, leaves, vegetables, and seeds
Breeding	2–4 babies

English Angora Rabbit

Angora rabbits are known for their long, soft hair. Angoras are coated with wool rather than fur, since they have no guard hairs. They are often bred commercially for their wool, which is removed by shearing or combing. Their coat can become uncomfortably matted, and since swallowed fur cannot leave their system, Angoras must be groomed at least every other day. There are many breeds of Angoras, including the French, German, giant, and satin. The English breed is popular as a pet because it is gentle and its face has something of a teddy bear expression. English Angoras are adorned with fluffy growths of wool even on their ears and face, except above the nose.

Scientific name	*Oryctolagus cuniculus*
Family	Leporidae
Size	2.3–3.4kg (5–7.5lb)
Distribution	Originated in the United Kingdom
Habitat	Domesticated
Diet	Grasses, leaves, vegetables, and seeds
Breeding	3–5 babies

English Silver Rabbit

Despite its name, this breed probably originated in France, although the rabbit had arrived in England as early as the beginning of the seventeenth century. During the eighteenth and nineteenth centuries, the English silver was being bred by the thousands for its fur and was used extensively in making hats. An alternative name for the breed was the Lincoln silver, as Lincoln was the center of England's hat industry. It took forty skins to make one fur hat. The present-day English silver, often referred to as the Argente Anglais, bears little resemblance to the nineteenth-century rabbit. Its short and glossy coat can be blue, fawn, cream, Havana, brown, or gray. The body colors are evenly silvered with white guard hairs throughout. The English silver is of medium size and is fine boned.

Scientific name	*Oryctolagus cuniculus*
Family	Leporidae
Size	2.7–4kg (6–9lb)
Distribution	Originated in France
Habitat	Domesticated
Diet	Grasses, leaves, vegetables, and seeds
Breeding	3–6 babies

Florida White Rabbit

The Florida white was developed in the United States in the 1960s by crossing albino Dutch, white Polish, and white New Zealand rabbits. Since the breed was developed initially as a laboratory animal, it is easy to care for and highly docile. Today the Florida white is a popular pet and show rabbit, frequently being awarded best in show. The rabbit's mid-length, thick, soft fur appears only in pure white, while its eyes are a striking bright pink. Weekly groomings are all that is called for, except during shedding season, when twice-weekly attention may be needed. The Florida white's body is about as long as it is wide, making this a highly compact and stocky animal. The head is rounded while the ears are erect and well furred.

Scientific name	*Oryctolagus cuniculus*
Family	Leporidae
Size	1.8–2.7kg (4–6lb)
Distribution	Originated in the United States
Habitat	Domesticated
Diet	Grasses, leaves, vegetables, and seeds
Breeding	8–10 babies

Giant Chinchilla Rabbit

Chinchilla rabbits were originally bred in France for their size and meat. The three breeds of chinchilla rabbits are American, standard, and giant. The standard chinchilla's coat is the same gray—merging into white and tipped with black—as the fur-producing rodent after which the breed is named. As the name suggests, the giant is by far the largest chinchilla, having been developed in the United States as a cross with the Flemish giant. The giant chinchilla grows very fast as a young rabbit and will weigh as much as six pounds when only two months old. The breed is usually kept commercially, but giants also make a gentle and docile pet. Some of their owners say they are like "big babies." With their erect ears and beautiful fur, they are an attractive breed.

Scientific name	*Oryctolagus cuniculus*
Family	Leporidae
Size	5.4–7.2kg (12–16lb)
Distribution	Originated in the United States
Habitat	Domesticated
Diet	Grasses, leaves, vegetables, and seeds
Breeding	7–10 babies

Harlequin Rabbit

The harlequin rabbit is easily recognized by its unique parti-colored pattern, like that of the Harlequin clown. There are two varieties of harlequins: the Japanese and the magpie. Japanese harlequins are orange and black in a split pattern down the front of the face and body. The ears should be of different colors and each side of the face should be the color of the opposite ear. The feet are also of alternating colors. The rabbit's sides are striped black and orange. Magpies are similarly patterned, in white and black. A medium-sized rabbit, the harlequin is wide-bodied. Its fur is short, soft, and easy to take care of. This is a relaxed breed, making the harlequin a suitable pet for a novice rabbit owner. Harlequins are fairly rare, due to the difficulties of breeding perfectly patterned animals.

Scientific name	*Oryctolagus cuniculus*
Family	Leporidae
Size	2.5–3.6kg (5.5–8lb)
Distribution	Originated in France; rare in the United States
Habitat	Domesticated
Diet	Grasses, leaves, vegetables, and seeds
Breeding	3–5 babies

Himalayan Rabbit

Believed to have originated in the region surrounding the Himalayas, this breed is kept in a larger number of countries than any other rabbit and has been known by a confusing multitude of names, including Russian, Chinese, Egyptian, African, Windsor, and black-nosed. The historical reason for the rabbit's widespread popularity was high demand for its pelt, an ersatz ermine. A Himalayan's fur is short, fine, and pure white, while it displays dark points on its legs, nose, ears, and tail. The original Himalayan variety had black points, but today lilac, chocolate, and blue points are recognized. These markings will not be fully apparent until the rabbit is mature. A Himalayan always displays bright pink eyes. This gentle-tempered breed has a slender, almost cylindrical body that makes it easy for children to handle.

Scientific name	*Oryctolagus cuniculus*
Family	Leporidae
Size	1–2kg (2.5–4.5lb)
Distribution	Originated in China or northern India
Habitat	Domesticated
Diet	Grasses, leaves, vegetables, and seeds
Breeding	2–4 babies

Holland Lop Rabbit

Any rabbit breed with ears that fall vertically is known as a lop. Named the miniature lop in the United Kingdom, the Holland lop is the smallest breed in the lop family. This rabbit had its beginnings in 1950, when a Dutch breeder decided to cross the large French lop with the Netherland dwarf. The resulting rabbit has short, bell-shaped ears. It is muscular and compact, with a massive head and short legs. The breed makes a popular pet due to its cute appearance and calm temperament, although this lop will become frightened if it is not handled carefully. When particularly pleased, the Holland lop may exhibit a characteristic display in which it jumps in the air and then runs around in circles. With its thick fur, this rabbit will require occasional grooming.

Scientific name	*Oryctolagus cuniculus*
Family	Leporidae
Size	1.4–1.8kg (3–4lb)
Distribution	Originated in the Netherlands
Habitat	Domesticated
Diet	Grasses, leaves, vegetables, and seeds
Breeding	2–4 babies

Mini Lop Rabbit

Also known as the German lop, the mini lop is the second-smallest of the lop-eared, or floppy-eared, rabbits, after the Holland lop. Larger breeds are the American fuzzy, English, French, plush, and velveteen lops. The mini lop was developed in Germany as a cross between larger German lops and chinchilla rabbits. The breed is easily recognized from its short, bulky body, giving the rabbit an almost ball-like appearance. The head is large and crowned by the eponymous thick, lopped ears, which hang close to the pronounced cheeks. The shape of the dangling ears resembles a horseshoe shape. The rabbit's thick, soft fur will need regular grooming to prevent fur balls. The most common coloration is agouti, but all colors are permissible, plus butterfly, harlequin, and black and tan patterns.

Scientific name	*Oryctolagus cuniculus*
Family	Leporidae
Size	2–3kg (4.5–6.5lb)
Distribution	Originated in Germany
Habitat	Domesticated
Diet	Grasses, leaves, vegetables, and seeds
Breeding	3–6 babies

Mini Rex Rabbit

The rex breed is also known as the velveteen due to its dense, short, and soft fur. Rex fur has shortened guard hairs, giving the coat its characteristic velvet feel. The two types of rex rabbits are the mini and the standard, with the mini being derived from the standard breed. The rex is renowned as a good breeder and a good mother. Rex rabbits are used as foster mothers for rabbits of other breeds. The mini rex is known for its skill in the sport of rabbit hopping, in which rabbits hop over a course of obstacles like show-jumping horses. The sport is popular in Europe and the United States. The mini rex has long, sharp, and rather scratchy toenails. These are curious rabbits, inquisitive and lively, and the mini rex in particular is a very popular pet.

Scientific name	*Oryctolagus cuniculus*
Family	Leporidae
Size	1.4–2kg (3–4.5lb)
Distribution	Originated in the United States
Habitat	Domesticated
Diet	Grasses, leaves, vegetables, and seeds
Breeding	9 or more babies

Netherland Dwarf Rabbit

Due to their small size and cute appearance, many of the rabbits sold in pet stores are Netherland dwarfs. This rabbit has a very large head, with wide eyes and a rounded face, giving it a babylike appearance throughout its life. Its small ears are carried high on its head. The breed was first developed in the Netherlands in the early years of the twentieth century by crossing Polish and native wild rabbits. Selective breeding has since encouraged gentleness and friendliness in the breed, but it retains a lively disposition inherited from its wild ancestors. Since these rabbits can be easily stressed, they are not the perfect choice for young children. They do not like to sit still for too long and have a tendency to nibble on anything within reach.

Scientific name	*Oryctolagus cuniculus*
Family	Leporidae
Size	1–2kg (2.5–4.5lb)
Distribution	Originated in the Netherlands
Habitat	Domesticated
Diet	Grasses, leaves, vegetables, and seeds
Breeding	2–4 babies

New Zealand Rabbit

Despite its name, the New Zealand rabbit was developed in the United States, although one story holds that the breed was taken to America by Kiwi sailors. The rabbit's ancestry includes Flemish giants, Angoras, and Americans, giving it a large size and dense fur. This is a quick-growing breed, and is kept commercially for its meat. The New Zealand is solid, with broad hindquarters. Its large head is crowned with round-tipped and furry ears. With its very thick, luxurious coat, this rabbit will need regular grooming. Officially recognized colors are rich shades of red, white, black, and blue, although blue is not accepted by the American Rabbit Breeders' Association. The New Zealand makes a docile pet on the whole, but adults are occasionally more aggressive than other breeds.

Scientific name	*Oryctolagus cuniculus*
Family	Leporidae
Size	4–5.5kg (9–12lb)
Distribution	Originated in the United States
Habitat	Domesticated
Diet	Grasses, leaves, vegetables, and seeds
Breeding	6–8 babies

Rhinelander Rabbit

As its name suggests, the Rhinelander was developed in Germany, at around the turn of the twentieth century. The Rhinelander is one of the few tricolored rabbits: it has a white body with black and bright orange spots. Its fur is short enough to snap back into place when you run your hand along it, and it is relatively easy to take care of, making it a good choice for a first-time owner. Its bright and bold eyes are brown in color. This is an arched breed, which means that its body arches from the shoulders through to the hips, giving it a graceful appearance. A nonaggressive rabbit, the Rhinelander is sometimes described as laid-back and is not known for excessive biting. Rhinelanders are great mothers and have large litters.

Scientific name	*Oryctolagus cuniculus*
Family	Leporidae
Size	3–4.5kg (6.5–10lb)
Distribution	Originated in Germany
Habitat	Domesticated
Diet	Grasses, leaves, vegetables, and seeds
Breeding	6–12 babies

Satin Rabbit

The satin is known for the high shine of its coat. This mutation was first noticed in a litter of Havana rabbits in the 1930s. When these rabbits were exhibited in the United States, there was such a furor that the satin rabbit was immediately recognized as a breed in its own right. Satin fur is distinguished by a translucent sheath to its guard hairs and needs no special care other than routine brushing. Satin rabbits are stocky and appear in a wide range of solid colors as well as Siamese and bicolor. The satin has a friendly disposition and seems to enjoy being petted. Since they are fairly large rabbits, satins will require a roomy cage. One drawback is that young children may have trouble holding a rabbit this bulky.

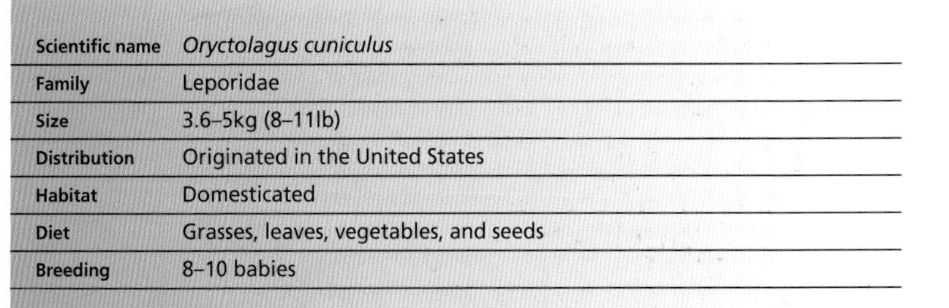

Scientific name	*Oryctolagus cuniculus*
Family	Leporidae
Size	3.6–5kg (8–11lb)
Distribution	Originated in the United States
Habitat	Domesticated
Diet	Grasses, leaves, vegetables, and seeds
Breeding	8–10 babies

Swiss Fox Rabbit

This beautiful long-haired rabbit was developed by crossing the Angora, chinchilla, and Havana breeds. The aim was to re-create the coat of the blue Arctic fox. The breed was very rare until recently, but it is now growing in popularity. The Swiss fox is a docile rabbit but is not an ideal pet for children, since its long coat is normally 4–7 centimetres (1.5–2.75 inches) long and is time-consuming to groom. If the hair is not cared for meticulously, the rabbit will become uncomfortably matted. Swiss foxes are strong and have somewhat round bodies. The neck is nearly nonexistent, and the legs, shoulders, and chest all well muscled. The muzzle is well defined and the 10-centimetre (4-inch) ears are upright. The Swiss fox is bred in white, blue, black, and Havana colors.

Scientific name	*Oryctolagus cuniculus*
Family	Leporidae
Size	2.5–3kg (5.5–7lb)
Distribution	Originated in Switzerland and Germany
Habitat	Domesticated
Diet	Grasses, leaves, vegetables, and seeds
Breeding	3–6 babies

Pallid Gerbil

Pallid gerbils are considered to be a good second species for those experienced in keeping the better-known Mongolian gerbil. Pallids are slender, with slightly protruding dark eyes, lengthy feet, and a tail longer than the body. They have a pale orange coat, with white on the belly and around the eyes. The pallid is fairly easy to care for thanks to its friendly temperament. It prefers to be kept with others of its species, so that they can sleep, groom, play, and engage in boxing matches together. Pallids do not like to keep still when being handled, but may enjoy climbing over their owner, provided they are protected from falls. When carefully looked after and properly housed in a transparent tank with a paper wool or hay bedding, a pallid can live for up to four years.

Scientific name	*Gerbillus perpallidus*
Family	Muridae
Size	8–13cm (3–5in) long, excluding tail
Distribution	Northwestern Egypt; illegal to own in California
Habitat	Rocky terrain and scrubland
Diet	Gerbil mix supplemented with crickets and mealworms
Breeding	4–8 young once a month

Mongolian Gerbil

Also known as the Mongolian jird, this species is the most popular member of the gerbil subfamily to be kept as a pet. Many different colors and coat patterns of this small rodent have been bred in captivity. The gerbil makes a good pet due to its hardiness and friendliness toward gentle humans. These are highly social animals and are happiest when kept in a pair, but when choosing your pets, be sure to purchase individuals of the same sex, preferably from the same litter. The best choice for housing is a large glass or transparent acrylic tank with a deep layer of corncob bedding, small animal substrate, peat, or aspen wood shavings. Your gerbils will also need an accessible water bottle and a regular sand bath to keep their coats clean.

Scientific name	*Meriones unguiculatus*
Family	Muridae
Size	10–15cm (4–6in) long, excluding tail
Distribution	Mongolia; illegal to own in California
Habitat	Semidesert and steppe
Diet	Gerbil mix supplemented with vegetables such as carrots
Breeding	Up to 12 young once a month

House Mouse

The domesticated mice kept as pets belong to the species *Mus musculus*. Selective breeding has created a range of colors not found in nature, in both solid colors and patterns. Mice have been popular pets for millennia because they are clean and small, and can usually come to enjoy responsible handling. Mice rarely bite except when frightened—but as they can be injured, young children should be supervised. Since mice are sociable animals, it is a good idea to keep them in groups, but if a male and female of breeding age are placed together, they will have a new litter every three weeks until separated. In their cage, mice will enjoy tubes and wheels. Outside the cage, an inquisitive mouse will enjoy exploring its surroundings, as well as hiding in dark corners.

Scientific name	*Mus musculus*
Family	Muridae
Size	15–30cm (6–12in) long, excluding tail
Distribution	Worldwide except Antarctica
Habitat	Close proximity to humans
Diet	Suitable feed mix, plus cereals and seeds
Breeding	4–12 young up to 15 times a year

Common Rat

The rats commonly kept as pets, often called the "fancy rats," are a domesticated breed of the common rat. Pet rats—whose behavior differs from their wild relatives depending on how many generations they are separated—are sociable and intelligent, and can even be trained to use a litter box. These pets can live for two to three years and come in a variety of different colors, coat types, markings, and eye colors. The rat is active at night and can emit and hear ultrasonic vocalizations. When pleased, rats may emit short bursts of ultrasonic chirps similar to laughing, although humans cannot hear this without special equipment. A group of rats will create a social hierarchy, with each rat knowing its place. Groups will commonly groom and sleep together.

Scientific name	*Rattus norvegicus*
Family	Muridae
Size	25cm (10in) long, excluding tail
Distribution	All continents except Antarctica; Canada and New Zealand have restrictions on rat ownership
Habitat	Wherever humans live, particularly urban areas
Diet	Omnivore, with a high cereal intake
Breeding	Average of 7 pups up to 5 times a year

Ferret

The ferret is a close relative of the polecat and a member of the weasel family. Believed to have been domesticated at least 2,500 years ago, the ferret was probably first used for hunting rabbits or other small animals. With its long, slim body and inquisitive nature, the ferret is ideal for chasing rodents out of their burrows, but this practice is illegal in several countries. Ferrets are crepuscular, which means they are most active at dusk and dawn. They spend up to eighteen hours a day sleeping but are extremely engaging, active, and energetic when awake. A ferret will commonly enjoy playing hide-and-seek with its owners. It is worth noting that ferrets do bite and should be monitored closely around young children. Due to a ferret's territorial nature, care should be taken when introducing new ferrets to a group.

Scientific name	*Mustela putorius furo*
Family	Mustelidae
Size	0.9–1.8kg (2–4lb); 50cm (20in) long, including tail
Distribution	Owning of ferrets is illegal or restricted in some countries and states, including California, Hawaii, Australia, and New Zealand
Habitat	Domesticated
Diet	Carnivore, low sugar, low fiber, and lactose-free
Breeding	6–8 kits in a litter; pet neutering is highly recommended

Sugar Glider

This marsupial is an increasingly popular pet in the United States. Pets must be obtained from a registered breeding program, and permits are required in some areas. Skin membranes extend from the fifth fingers of the glider's forelimbs to the first toes of its hind legs, allowing the marsupial to sail from tree to tree for distances of up to 50 metres (55 yards). In the wild, the sugar glider is active by night, when it feeds on the sap and pollen of eucalyptus, acacia, and gum trees and hunts for insects and other small animals. With plenty of human attention, and preferably the company of its own species, a glider will make a lively and entertaining pet. A large cage with plenty of toys and branches for exercise will be needed.

Scientific name	*Petaurus breviceps*
Family	Petauridae
Size	85–140g (3–5oz); 15–20cm (6–8in) long
Distribution	Eastern and northern Australia and New Guinea; illegal to own in some countries and states
Habitat	Among hollow trees
Diet	Suitable mixes containing honey and cereal, insectivore feeds, and fruit
Breeding	2 young once a year

Chinese Water Dragon

This active lizard is green with a vertebral crest and green and dark brown tail bands. Lighter diagonal stripes of green or turquoise may be visible around the body. The throat is orange or yellow, in a single color or striped. The whiplike tail can be used as a weapon, although this species makes a relatively easygoing pet. In the wild, this tree-dwelling lizard lives close to freshwater. When threatened, it can drop from its perch into water to swim for safety or remain underwater for up to thirty minutes. In captivity, the Chinese water dragon needs a large enclosure, kept humid and at 29–35°C (85–95°F) during the day. There should be branches for perching and a large pool for submersion. The basking area needs to be warmed by a UVA–UVB light.

Scientific name	*Physignathus cocincinus*
Family	Agamidae
Size	60–90cm (24–36in) long
Distribution	India, China, and Southeast Asia
Habitat	Forests; on the banks of lakes and streams
Diet	Insects, earthworms, and fish, plus some vegetables, fruit, and mice
Breeding	2–4 eggs

Bearded Dragon

This large lizard earns its common name with its spiny throat. When threatened, the lizard puffs out this "beard" while also opening its jaws to frighten off predators. Males have a darker beard than females, and it darkens further when courting. In the wild, these lizards spend the hottest part of the day hiding from the sun in a cool spot or underground, and bask in the mornings and evenings. In a group of bearded dragons, the most dominant individuals will occupy the best basking positions, often clambering over others to get there. Captive-bred bearded dragons are commonly available in a wide variety of colors, including brown, gray, rust, orange, and green. A 430-litre (95-gallon) terrarium will be necessary to house just one adult, with pairs and groups requiring significantly more room.

Scientific name	*Pogona vitticeps*
Family	Agamidae
Size	40–60cm (16–24in) long
Distribution	Central Australia
Habitat	Dry woodland to rocky desert
Diet	Insects, vegetables, and greens
Breeding	11–30 eggs

Painted Terrapin

Sadly, the painted terrapin is a critically endangered species, which means that it is close to extinction, so ensure that any pets you obtain are captive-bred. There are perhaps fewer than 2,000 painted terrapins left in the wild, so any suspected trade in wild animals should be reported. The carapace of this large, river-dwelling terrapin is gray-brown and features a bony ridge running from the head to the tail. The painted terrapin's upturned nose is often said to resemble a snorkel. In the wild, these terrapins live in tidal rivers, feeding on the vegetation and fruits on their banks. When nesting, females move to sandy areas a few miles from the river. Nesting takes place at night, and hatchlings will emerge about ten weeks later. Painted terrapins can be kept in a large aquarium with ample basking areas.

Scientific name	*Callagur borneoensis*
Family	Bataguridae
Size	55–60cm (22–24in) long
Distribution	Brunei, Indonesia, Malaysia, and Thailand
Habitat	Tidal rivers
Diet	Greens, feeder fish, and insects
Breeding	20 eggs

Emerald Tree Boa

The adult of this species is brilliant green with a yellow underside. White bars dot the body. Newborns are red or yellow and will achieve the adult coloration by about nine months old. This snake can deliver a painful bite and is hard to tame, so is not for novice hobbyists or for those hoping to handle their pet. Since these snakes are tree dwellers, they need a tall terrarium with plenty of sturdy branches for climbing. However, these snakes are fairly inactive and may appear to be motionlessly coiled around a tree branch for weeks at a time. The terrarium should have a daytime temperature of about 29°C (85°F), with a humidity of more than 85 per cent, which can be achieved only by misting several times a day. A solitary species, the emerald tree boa is best housed individually.

Scientific name	*Corallus caninus*
Family	Boidae
Size	1.8–2m (6–7ft) long
Distribution	South America: Amazonia
Habitat	Rain forest
Diet	Mice, rats, and chicks
Breeding	7–14 live young

Rainbow Boa

There are nine subspecies of rainbow boas, all of which are known for their iridescent sheen, particularly in sunlight. These nonvenomous constrictors are nocturnal and, like the emerald tree boa, locate their prey using heat-sensitive pits in the scales of their lips and snout. Rainbow boas grow to be very large and require careful monitoring of the humidity and temperature in their terrarium. Young pets are likely to bite frequently but may become more docile as they accustom themselves to handling. As a result, rainbow boas are not suited to complete novices. Although rarer in captivity than some of its relatives, the Colombian rainbow boa (***Epicrates cenchria maurus***) is known for being one of the more docile subspecies. It is tan colored with a faint pattern of circles and spots.

Scientific name	*Epicrates cenchria*
Family	Boidae
Size	1–1.4m (3.5–4.5ft) long
Distribution	Central and South America
Habitat	Forest and woodland
Diet	Chicks, mice, and rats
Breeding	10–30 live young

Veiled Chameleon

These striking chameleons have an extravagant decorative growth called a casque on their heads. Males are green, varying from lime to olive depending on mood, health, and temperature. They are marked with spots and stripes of yellow, brown, and blue. Females and juveniles are green with white markings. Breeding females are a very dark green with blue and yellow spots. Veiled chameleons are tree dwellers and have a long, sticky tongue to capture their insect prey. A large and airy wire cage—filled with leafy branches and potted plants—is necessary for keeping them. The leaves should be misted twice a day to enable the chameleons to drink, as they prefer to drink water that is in drops. Male veiled chameleons must always be kept out of sight of other males.

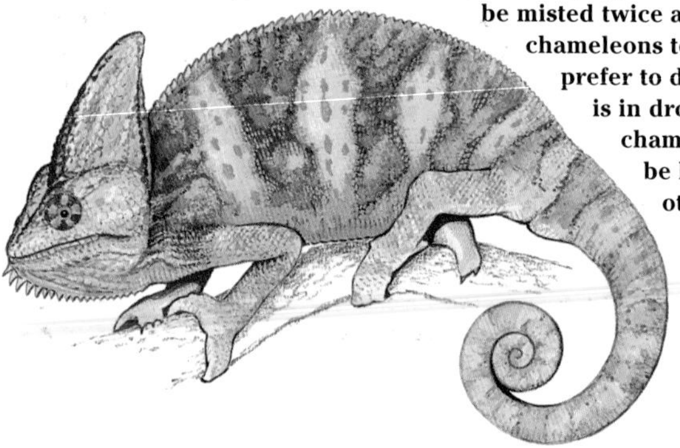

Scientific name	*Chamaeleo calyptratus*
Family	Chamaeleonidae
Size	30–60cm (12–24in) long
Distribution	Yemen and Saudi Arabia
Habitat	Trees in mountainous regions
Diet	Insects such as crickets, mealworms, and waxworms, plus some blossoms and leaves
Breeding	20–70 eggs up to 3 times a year

Jackson's Chameleon

Male Jackson's Chameleons have two slightly upturned horns above their eyes and another horn on their nose, giving this species its alternative common name of the three-horned chameleon. Females have either very small horns or lack them entirely. This species is bright green, while some individuals show traces of blue and yellow. Coloring will change quickly depending on temperature, health, and mood. Males are territorial and should not be placed in a cage with other males. A male and one or two females will do well in a very large wire cage with leaves and branches to provide visual barriers. Males may live for eight or more years, while females have shorter life spans. Unlike most chameleons, which lay eggs, females of this species give birth to live young, after a five- to six-month gestation period.

Scientific name	*Chamaeleo jacksonii*
Family	Chameleonidae
Size	30–38cm (12–15in) long
Distribution	Kenya and Tanzania; introduced to Hawaii
Habitat	Montane forests over 10,000 ft.
Diet	Grasshoppers, crickets, mealworms, silkworms, and waxworms
Breeding	8–30 live young

Common Snake-Necked Turtle

As its name suggests, this species of turtle has an extremely lengthy neck, which can be as long as its carapace. It is a side-necked turtle, which means that it bends its neck sideways into its shell rather than pulling it directly back. The carapace is flattened, broad, and colored brown. This turtle's powerful webbed feet are ideal for both swimming and tearing apart its prey. In the wild, these mainly aquatic turtles bask on rocks and logs during the day. Also known as the "stinker," this turtle gives off a strong smell from its scent glands when it is threatened. This is one of the most popular and easy-to-keep pet turtles, and it adapts well to life in a home aquarium. Snake-necked turtles need a large aquarium with basking space.

Scientific name	*Chelodina longicollis*
Family	Chelidae
Size	40–46cm (16–18in) long, including neck
Distribution	Eastern Australia: northern Queensland to Victoria
Habitat	Swamps, dams, and lakes
Diet	Minnows, worms, insects, and trout or catfish pellets
Breeding	8–24 eggs

Victoria Short-Necked Turtle

The Victoria species is a member of the Australian short-necked turtles *Emydura* genus. The six species of the genus are web-footed and semiaquatic river turtles with unusually short necks. This species is rare outside the Australian pet trade. All pets for sale must be captive-bred, since Australia bans the catching and exporting of wild specimens. The Victoria turtle has an oval, domed carapace, which is olive-brown to black with dark blotches. The head is broad with a slightly projecting snout and is a brown or greenish gray, like the limbs. On each side of the head are two salmon-colored stripes. Males have longer and thicker tails than females. An aquarium for two adults must have at least a 270-litre (60-gallon) capacity and feature a sloping ramp that allows the turtles to exit the water to bask.

Scientific name	*Emydura victoriae*
Family	Chelidae
Size	25–30cm (10–12in) long
Distribution	Northwestern Australia to the vicinity of Darwin, Northern Territory, and border areas of northwestern Queensland
Habitat	Large rivers and streams; large water holes on floodplains
Diet	Shellfish, minnows, worms, insects, and trout or catfish pellets
Breeding	7–14 eggs

Hilaire's Side-Necked Turtle

This large and hardy turtle is primarily a water dweller. A pair will require at least an 800-litre (180-gallon) filtered aquarium with a ramp for exiting the water to bask. In warm climates, some owners put these turtles in a fenced pond outdoors. The Hilaire's turtle has an oval, flattened carapace with parallel sides. The carapace is dark brown, gray, or olive, with a yellow border. The head is large and broad with a projecting snout and two chin barbels. A pronounced black stripe runs along each side of the head, which is gray to olive above and creamish white below the stripe. In the wild, eggs are laid in nests dug in beaches. This species likes to avoid the sand near the water when digging a nest, preferring higher and harder ground.

Scientific name	*Phrynops hilarii*
Family	Chelidae
Size	34–40cm (13.5–16in) long
Distribution	South America: Southern Brazil, Uruguay, northern Argentina, and Paraguay
Habitat	Ponds, lakes, and swamps with abundant vegetation
Diet	Snails, fish, and freshwater turtle pellets
Breeding	15–20 eggs

Mangrove Snake

The mangrove snake is also known as the gold-ringed cat snake, for the yellow bands that mark its black body. There are nine similar-looking subspecies; only the subspecies from Sulawesi in Indonesia (*Boiga dendrophila gemmicincta*) is completely black when adult. The mangrove snake is the largest in the cat snake family, which consists of arboreal, nocturnal snakes with long, slim bodies. The mangrove snake is mildly venomous but is not life-threatening to humans. It is a popular snake to keep as a pet because it is relatively hardy. Most specimens available in the pet trade will have been caught in the wild. After they have settled into the home, they are adaptable and easygoing. In the wild, these snakes eat lizards, birds, frogs, other snakes, and rodents. Accustoming them to a rodent-only diet can be tricky for a novice.

Scientific name	*Boiga dendrophila*
Family	Colubridae
Size	1.8–2.4m (6–8ft) long
Distribution	India and Southeast Asia
Habitat	Lowland rain forests, often in mangrove swamps
Diet	Mice and rats
Breeding	4–15 eggs

African Egg-Eating Snake

The five species and numerous subspecies of African egg-eaters display a range of patterns and colorations, from browns, greens, and reds to black. *Dasypeltis scabra* and *D. medici* are the species most often seen in the pet trade. These nonvenomous snakes have adapted to feeding exclusively on eggs and have an extraordinary method of consuming eggs much larger than their own head. These excellent tree climbers home in on suitable eggs by scent. They wrap their toothless mouth around the egg, then flex their muscles, cracking the egg on bony projections on their backbone. Every last bit of liquid is squeezed out before the crushed shell is regurgitated. In captivity, these snakes will do well kept in a pair or trio in a large terrarium with some opportunities for climbing.

Scientific name	*Dasypeltis* genus
Family	Colubridae
Size	0.3–1m (1–3.5ft) long
Distribution	Africa
Habitat	Forest to grassland
Diet	Small hen and quail eggs
Breeding	6–25 eggs

Beauty Rat Snake

A slender, semiarboreal snake, the beauty rat snake usually has a ground color of cream to greenish-yellow. The back of most subspecies is marked with a repeating pattern of two black rounded spots, which merge in the center. Subspecies vary in their body patterning and coloration, including *Elpha taeniura grabowskyi* (pictured), but most have a dark line passing through the eye from the snout to the jaw. Beauty rat snakes caught in the wild can continue to be temperamental and keen to bite many years after capture. They may also be infested with parasites. Try to obtain a captive-bred specimen. Carefully kept individuals can live for up to fifteen years. Since this snake likes to climb, it is best to provide a large aquarium, turned on end to create a vertical enclosure, and filled with plenty of stout branches.

Scientific name	*Elaphe taeniura*
Family	Colubridae
Size	1.5–2m (5–7ft) long
Distribution	Asia: China, Taiwan, Burma, and Thailand
Habitat	Tropical forests
Diet	Mice, rats, and chicks
Breeding	9–20 eggs

Sungazer Lizard

The sungazer, also known as the giant girdled lizard, is a well-armored, yellowish-brown lizard. It is protected by great spines on the back of the head, as well as spiny scales all over its body and tail. The tail is particularly effective when flicked at predators. This lizard gets its common name from the habit of sitting outside its burrow basking in the sunshine of its South African homeland. The sungazer is classed as vulnerable in the wild, but captive-bred individuals are occasionally available, commanding a consequently high price. A group of sungazers, with no more than one male, will do well in a terrarium of at least 220 litres (48 gallons) with a deep sandy substrate. A UVA and UVB heat lamp should be placed at one end of the enclosure for basking.

Scientific name	*Cordylus giganteus*
Family	Cordylidae
Size	15–18cm (6–7in) long
Distribution	South Africa
Habitat	Grassland
Diet	Insects and chopped fruit
Breeding	1–2 live young; rarely breeds in captivity

Brown Basilisk Lizard

Sometimes known as "Jesus lizards," basilisk lizards can seemingly walk on water. When moving fast to escape a predator, their webbed feet allow them to run across water for up to 20 metres (65 feet) without sinking. This species is brown in color, and males have a high head crest. A large cage with a capacity of 270–430 litres (60–95 gallons) is needed, with plenty of planted visual barriers to prevent these active and nervous lizards from picking up momentum and running into the glass or wire of the cage edges. Set up a humid aquarium with a large water receptacle, along with roomy perches. Brown basilisks can live for up to ten years in captivity. Males will fight with other males, but a mixed group consisting of one male and one or more females will get along well.

Scientific name	*Basiliscus vittatus*
Family	Corytophanidae
Size	25–60cm (10–24in) long
Distribution	Central America: Mexico to Colombia; recently introduced to Florida
Habitat	Close to bodies of water
Diet	Insects plus some blossoms
Breeding	2–18 eggs

Painted Turtle

There are four subspecies of the painted turtle: the western, eastern, southern, and midland. All subspecies have olive to black carapaces with red on the marginal scutes and yellow or red stripes or spots on their heads. The head stripes and the beautiful "painted" patterns of the shell's plastron, or underside, differ between subspecies. Painted turtles are most active from March to October. During the winter, they hibernate by burying themselves in the mud beneath streams and lakes, which insulates them from the cold. In this way, these turtles can survive without oxygen longer than any other known air-breathing vertebrate. Up to four adults can be housed in a 230-litre (50-gallon) aquarium or fenced garden pond. As long as their water is kept clean, these hardy turtles can survive for as long as twenty-five years.

Scientific name	*Chrysemys picta*
Family	Emydidae
Size	11–25cm (4.5–10in) long
Distribution	Southern Canada, United States, and northern Mexico
Habitat	Ponds, lakes, marshes, and slow-moving rivers with muddy bottoms
Diet	Insects, fruits, greens, trout or catfish pellets, and turtle foods
Breeding	4–15 eggs

Red-Eared Slider

These popular turtles have a distinctive red stripe along the sides of their heads. In the United States, members of this genus are usually called turtles, but in the United Kingdom they are split into turtles (aquatic), tortoises (land-dwelling), and—like the red-eared slider—terrapins, which are semiaquatic. Red-eared sliders leave the water to bask and lay eggs. They can live for more than fifty years in captivity. A large aquarium is needed to house red-eared sliders, with just one adult needing a 270-litre (60-gallon) capacity. Deep, filtered water is required for swimming alongside a basking area warmed and illuminated by a UV bulb. In warm summer weather, a fenced outdoor pond would be enjoyed. Red-eared sliders can become aggressive and may injure or even kill other turtles if kept together.

Scientific name	*Trachemys scripta*
Family	Emydidae
Size	20–30cm (8–12in) long
Distribution	Southern United States: Mississippi River to the Gulf of Mexico; certain areas have restrictions on turtle ownership
Habitat	Ponds, marshes, and streams with basking areas
Diet	Greens, fruits, insects, and prepared turtle foods
Breeding	2–30 eggs, with bigger females laying larger clutches

Green Iguana

The green iguana usually grows to 1.5 metres (5 feet) or more. It is a bright green lizard, with a tail banded with black. Adults have a dewlap, a flap of skin hanging below the jaw. There is a row of spines along the back and tail, which helps to protect this iguana from predators. Before adopting a green iguana, consider that zoos are inundated with adults that have become too large and aggressive for their owners to handle. Babies may look like a manageable size, but they soon outgrow a standard aquarium. Green iguanas given the run of the house frequently sustain injuries and cause damage. There are very few owners with the experience or resources to adequately house or care for one of these reptiles.

Scientific name	*Iguana iguana*
Family	Iguanidae
Size	1.4–2m (4.5–6.5ft) long
Distribution	Central and South America: Mexico to Brazil, Paraguay, and the Caribbean
Habitat	Forests close to water
Diet	Greens plus some vegetables and fruits
Breeding	20–70 eggs

Carolina Green Anole

Despite its color-changing abilities, this tree-dwelling lizard is not a true chameleon, being more closely related to the green iguana. The green anole's body color varies from green to brown, depending on temperature, surroundings, and mood. Both males and females sport a dewlap beneath the chin, with the male's being much larger and more colorful, in a pink to reddish shade. When engaging in a courtship or territorial display, males inflate the dewlap (pictured) and bob up and down as if dancing. When purchasing a green anole, beware of stores that keep overcrowded, stressed, and dehydrated specimens. Males are territorial and are best housed in a group with up to three females. A large, planted aquarium with plenty of vertical space for climbing should be provided.

Scientific name	*Anolis carolinensis*
Family	Polychrotidae
Size	10–24cm (4–9.5in) long
Distribution	United States: North Carolina to Texas; Caribbean
Habitat	Woodland to suburban areas
Diet	Calcium-dusted crickets and other small insects
Breeding	Up to 10 eggs laid in a season

Green Tree Python

Adults of this species are usually bright green with white dots or a thin white dorsal line. Yellow and turquoise individuals are also sometimes found. Hatchlings are yellow, orange, or red, taking on their adult coloration within three months. This nonvenomous arboreal snake spends much of its time resting in a tree, with a couple of coils looped over a branch in a saddle shape and its head in the middle (pictured). Prey is captured by striking out while dangling from a branch by the prehensile, or strongly gripping, tail. To house this snake, a 110-litre (25-gallon) or larger terrarium will be needed, turned on end to provide room for climbing in plenty of sturdy branches. Green tree pythons can be bad-tempered and should always be housed singly and handled with care.

Scientific name	*Morelia viridis*
Family	Pythonidae
Size	1.4–1.8m (4.5–6ft) long
Distribution	Cape York Peninsula of northeastern Australia, New Guinea, and Indonesia
Habitat	Rain forest, bush, and scrub
Diet	Chicks, mice, and rats
Breeding	12–25 eggs

Short-Tailed Python

There are three similar subspecies of this nonvenomous python, with the usual common names of Borneo short-tailed python, the red blood python, and the black blood python. Adults grow to up to 1.8 metres (6 feet) and have a wide girth. As the name suggests, this species' tail is very short in relation to the reptile's length. The python sports a pattern of chocolate to red-brown blotches on a ground of beige to gray-brown. Individuals caught in the wild tend to be more aggressive and unpredictable than captive-bred specimens. A large and fairly square terrarium would be preferred, with room for the snake to stretch out to its full length and to maneuver comfortably. The enclosure should be kept humid and well ventilated to avoid respiratory problems and to ensure that the python does not have difficulty shedding.

Scientific name	*Python curtus*
Family	Pythonidae
Size	1.5–1.8m (5–6ft) long
Distribution	Southeast Asia: Thailand, Malaysia, and Indonesia
Habitat	Marshes, swamps, and riverbanks in rain forests
Diet	Mice and rats
Breeding	10–12 eggs

Royal Python

The smallest of the African pythons, this snake makes a very popular pet. It is a stocky, nonvenomous snake with a pattern of light brown to green blotches on a black ground. Captive-bred populations have also developed a wide number of different colors and patterns. The name royal python is drawn from the legend that Cleopatra wore this snake on her arm. The python's other commonly used name, the ball python, is due to the fact that this snake curls into a ball when threatened, with its head and neck safely tucked away in the middle. Young captive snakes may be aggressive before they get used to human contact. This species is renowned for being a difficult feeder. Live rodents should never be left unattended in a snake's cage because a reluctant feeder can receive a deadly bite.

Scientific name	*Python regius*
Family	Pythonidae
Size	0.9–1.2m (3–4ft) long
Distribution	Africa: Senegal to Uganda
Habitat	Grassland
Diet	Mice and rats
Breeding	4–6 eggs

Common Sandfish Skink

This skink is best known for its unusual locomotion through desert sands. When threatened, overheated, or hunting, the skink dives into soft sand dunes, propelling itself in a manner similar to swimming. It can home in on the vibrations made by sand-burrowing insects with deadly accuracy. An attractive skink, the common sandfish has a creamy-yellow coloration with well-defined brown-black cross bands. Its body shape is highly streamlined. In captivity, the common sandfish can live for up to ten years in optimum conditions. A group of up to three lizards will need a 110-litre (25-gallon) terrarium. A deep, fine, sandy substrate is necessary, as these skinks will spend much of their time burrowing. A heat lamp should warm the sand at one end of the tank to 49°C (120°F).

Scientific name	*Scincus scincus*
Family	Scincidae
Size	13–20cm (5–8in) long
Distribution	North Africa to the Middle East
Habitat	Sandy desert
Diet	Insects and chopped fruit
Breeding	5–10 eggs

Radiated Tortoise

These beautiful tortoises are extremely endangered due to loss of habitat and overexploitation by the pet trade. If you are willing to support a captive-breeding program and put in the time and money to locate a captive-bred specimen, complete with a permit, these rare tortoises can make extremely rewarding pets. Any suspected trade in wild-caught specimens should be reported. The radiated tortoise's name comes from the striking yellow lines that radiate from the center of each plate on its shell. It has yellowish feet, tail, and head. Males have longer tails than females. These tortoises do not do well in cold weather and should not be kept in areas where temperatures fall below 16°C (60°F) at night. If well kept, radiated tortoises can live as long as fifty years.

Scientific name	*Astrochelys radiata*
Family	Testudinidae
Size	30–40cm (12–16in) long
Distribution	South and southwest Madagascar
Habitat	Dry regions of brush and woodland
Diet	Grasses, dark green leafy vegetables, and small amounts of other vegetables and fruit
Breeding	3–12 eggs

Hermann's Tortoise

Like most tortoises, these reptiles live on dry land. A small- to medium-sized tortoise, the Hermann's can live for twenty-five to thirty-five years or more. Its carapace, or shell, is highly domed. Young animals have black-and-yellow patterned carapaces, usually fading with age to a less distinct gray or yellow. There are two subspecies of Hermann's tortoises: the western and the eastern, with the western being much smaller. In temperate climates, these tortoises should be kept in a large outdoor pen in the warm summer months, with a house for shelter. For the rest of the year, a roomy indoor terrarium should be offered, with a UV bulb for basking. These tortoises hibernate in winter, with adults sleeping for four to five months. A secure box should be provided, with the temperature being maintained at about 5°C (41°F).

Scientific name	*Testudo hermanni hermanni* (western) and *Testudo hermanni boettgeri* (eastern)
Family	Testudinidae
Size	Western 6–19cm (3–7.5in) long; eastern 14–28cm (5.5–11in) long
Distribution	Southern Europe: eastern Spain to Albania
Habitat	Meadows
Diet	Greens with some fruit and a small amount of animal protein
Breeding	5–10 eggs

Arabian Horse

The ancestors of the Arabian horse were probably first domesticated by Bedouin nomads living in the Arabian Peninsula, perhaps as much as 4,000 years ago. Arabians were bred as warhorses, exhibiting endurance, speed, and intelligence, with beauty and elegance also being prime requirements. The modern-day Arabian has a refined, wedge-shaped head, with a distinctive concave, or dished, profile, and a high tail carriage. Coat colors are bay, chestnut, black, gray, and roan. Arabians that appear white are nearly always actually grays. With its great gentleness and sensitivity, this hotblood responds quickly to its rider, but will become highly strung if not treated with corresponding intelligence and respect. Arabians are widely used for pleasure riding, trail riding, and competitions in a wide range of disciplines, from racing and youth events to endurance riding.

Scientific name	*Equus caballus*
Family	Equidae
Size	14.1–15.1 hands high; 1.43–1.53m (4.8–5ft) tall
Distribution	Originated in the Arabian Peninsula
Habitat	Domesticated
Diet	Forage from hay or pasture, plus some grain
Breeding	Usually one foal

Mongolian Horse

The native horse breed of Mongolia, this horse is adapted to the extreme climate of the steppes. It is a small and stocky horse, with a relatively large head. Tests have shown that this breed has a wide genetic variety, suggesting that it is a very old breed that has developed with little human selection. Some enthusiasts claim that the Mongolian has remained unchanged since the days of Genghis Khan. Mongolians are very robust and can roam outdoors all year, foraging for their own food throughout the freezing Mongolian winters and scorching summers. They are used for all the work of the nomads who keep them, and also for racing. Once a Mongolian is familiar with carrying a rider, it is dependable, calm, and affectionate. Many other, more commonly available breeds descend from the Mongolian.

Scientific name	*Equus caballus*
Family	Equidae
Size	12–14 hands high; 1.22–1.42m (4–4.5ft) tall
Distribution	Originated in Mongolia
Habitat	Steppes
Diet	Forage from hay or pasture, plus some grain
Breeding	Usually one foal

Orlov Horse

Also known as the Orlov trotter, this famous Russian horse was bred for its fast trot and stamina. The horse is named after Count Alexei Orlov, who developed the breed in the late eighteenth century by crossing various European mares with Arabian stallions. The Orlov was popular for harness racing in nineteenth-century Russia, but fell from favor when the faster American Standardbred was imported. It looked as if the breed might disappear until, in 1997, the International Committee for the Protection of the Orlov Trotter was established. Today, twelve stud farms in Russia and three in Ukraine breed Orlov horses, which are appreciated worldwide as pleasure mounts as well as light draft horses. An Orlov has a large head; a long, arched neck; a well-muscled body; and powerful legs.

Scientific name	*Equus caballus*
Family	Equidae
Size	15.2–16.2 hands high; 1.54–1.65m (5.2–5.5ft) tall
Distribution	Originated in Russia
Habitat	Domesticated
Diet	Forage from hay or pasture, plus some grain
Breeding	Usually one foal

Trakehner Horse

Held to be the lightest and most refined of the warmbloods, the Trakehner has a great deal of Thoroughbred, Anglo-Arabian, and Arabian in its ancestry. As a result, some Trakehners can also be more spirited than the average warmblood. In general, Trakehners have good stamina and are agile and highly trainable. They make prized dressage mounts, while some individuals are world-renowned jumpers and eventers. The Trakehner displays its Thoroughbred blood with its fine head, well-set neck, sloping shoulders, strong back, and powerful hindquarters. Any coat color is permissible, with bay, chestnut, gray, and black being the most usual. The breed takes its name from the site of the main stud farm, Trakehnen, set up in 1732 by Frederick William I of Prussia. Today the town is called Yasnaya Polyana and is found in Russia.

Scientific name	*Equus caballus*
Family	Equidae
Size	15.2–17 hands high; 1.54–1.73m (5.1–5.7ft) tall
Distribution	Originated in Russia
Habitat	Domesticated
Diet	Forage from hay or pasture, plus some grain
Breeding	Usually one foal

Lipizzan Horse

This renowned horse originated in sixteenth-century Austria when the Hapsburgs brought Andalusian horses from Spain to establish studs. The Andalusians were bred with the native Karst horse, as well as the now-extinct Neapolitan, to develop horses for the increasingly popular Austrian riding schools. One of the studs, at Lipizza (now Lipica) in modern-day Slovenia, specialized in breeding horses for the Spanish Riding School in Vienna, named after the heritage of its horses. The school still demonstrates haute école dressage today, including the extraordinary stylized jumps known as "airs above the ground." In addition to excelling in dressage, Lipizzans are used for leisure riding around the world. Most Lipizzans are gray, with a long head, sturdy neck, wide chest, and muscular shoulders. The legs are strong, with broad joints and small, tough feet.

Scientific name	*Equus caballus*
Family	Equidae
Size	15–16.1 hands high; 1.52–1.64m (5–5.4ft) tall
Distribution	Originated in Austria and Slovenia
Habitat	Domesticated
Diet	Forage from hay or pasture, plus some grain
Breeding	Usually one foal

Kladruber Horse

With only about a hundred mares in existence, the Kladruber is a very rare breed. The horse has been bred on the Kladruby nad Labem stud farm in the Czech Republic for more than four centuries. The Kladruber breed and stud farm are designated as Czech historical landmarks. Used as draft horses or in the sport of combined driving, these horses are strong and muscular, with wide chests and hindquarters; long, swanlike necks; and powerful legs. The nose is prominent and Roman, while the eyes are large and lively. Today, Kladrubers are gray (pictured) or black, although in the past they have been a variety of colors. Most famously, Kladrubers serve on ceremonial occasions of the Danish royal court and are also used by the Swedish mounted police.

Scientific name	*Equus caballus*
Family	Equidae
Size	15.3–17.3 hands high; 1.55–1.76m (5.2–6ft) tall
Distribution	Originated in the Czech Republic
Habitat	Domesticated
Diet	Forage from hay or pasture, plus some grain
Breeding	Usually one foal

Knabstrup Horse and Pony

The Knabstrup is known for its spotted coat patterns, which range from the highly prized full leopard spotted (white with black, bay, or chestnut spots) to almost solid color. The Knabstrup is bred in four sizes and types: the sport horse, the classic, the pony, and the mini pony. The attractive pony, with its friendly temperament and good health, is a very popular choice for children. All the types are highly trainable and show good stamina, excelling at a variety of under-saddle disciplines. The breed is descended from the spotted horses of Spain and was established in 1812 when a chestnut blanket mare (with spotted white over the hip area) was bought from a Spanish cavalry officer and bred with native stallions in Knabstrupgaard in Nordsealand, Denmark.

Scientific name	*Equus caballus*
Family	Equidae
Size	Horse: 15.2–16 hands high; 1.54–1.62m (5.2–5.3ft) tall Pony: 10–13.3 hands high; 1–1.35m (3.3–4.6ft) tall
Distribution	Denmark
Habitat	Domesticated
Diet	Forage from hay or pasture, plus some grain
Breeding	Usually one foal

Selle Français Horse

The Cheval de Selle Français ("French saddle horse") is a popular sporting horse originating in France. This is a warmblood, which means that it is a middleweight horse bred from both hotbloods (such as Thoroughbreds and Arabians) and coldbloods (such as the Ardennes draft horse). The Selle Français stud book covers a wide range of types suited to different roles, but all of the breed are strong, agile, fast, and highly trainable. The sport horses particularly excel at show jumping, with numerous Selles Françaises having been the winners of the World Cup of Show Jumping or gold medalists in the World Equestrian Games. The nonspecialist horses make popular leisure horses and often put in an appearance at riding schools. Any coat color is permissible, but the most usual are bay or chestnut.

Scientific name	*Equus caballus*
Family	Equidae
Size	16–16.2 hands high; 1.62–1.65m (5.3–5.5ft) tall
Distribution	Originated in France
Habitat	Domesticated
Diet	Forage from hay or pasture, plus some grain
Breeding	Usually one foal

Ardennes Horse

Also known as the Ardennais, this is one of the oldest breeds of draft horses. Ardennes are heavy-boned, with short, thick legs; broad, compact bodies; strong heads; and a straight profile. They can be easily recognized by their feathered fetlocks. These horses have a proud and purposeful carriage, while their large, docile eyes have an intelligent expression. Coat colors are palomino, gray, chestnut, bay, brown, and roan. These powerful horses can weigh up to 1,000 kilogrammes (2,200 pounds). Ardennes are known for their calm and easygoing temperament and their suitability for working in the hilly terrain of their home region, the Ardennes plateau. The breed's endurance was valued by Napoléon Bonaparte, who used them in his infamous Russian campaign. Today, they are mainly used for drafting and farmwork.

Scientific name	*Equus caballus*
Family	Equidae
Size	15–16 hands high; 1.52–1.62m (5–5.3ft) tall
Distribution	Originated in the Ardennes region of France, Belgium, and Luxembourg
Habitat	Domesticated
Diet	Forage from hay or pasture, plus some grain
Breeding	Usually one foal

Camargue Horse

The world-famous Camargue horses live wild in the wetlands of the Rhône delta in southern France, with their ancestry dating back at least as far as the Roman invasion. The modern Camargue horse is hardy and relatively small, not reaching more than 14.2 hands high. Its square, heavy head points to its primitive ancestry, but the signs of Arabian and Thoroughbred blood can also be seen. All adult Camargues are gray, with black skin beneath a white coat. Traditional Camargue cowboys, known as *gardians*, look after the stock, rounding them up for yearly inspections. Only a foal born outdoors within the Camargue region can be registered as a true *sous berceau* ("within the birthplace") Camargue. The docility, stamina, and agility of the Camargues have led to their use in dressage and long-distance riding.

Scientific name	*Equus caballus*
Family	Equidae
Size	13.2–14.2 hands high; 1.34–1.44m (4.5–4.8ft) tall
Distribution	Camargue region of France; also bred in the United Kingdom
Habitat	Wetlands
Diet	Forage from pasture, plus some grain
Breeding	Usually one foal

Oldenburg Horse

Originating in the former grand duchy of Oldenburg in northwestern Germany, the Oldenburg is a sport horse with good jumping ability. Its warm-blooded heritage leaves the breed calm, highly rideable, and healthy, making it ideal for most leisure riders. As with all warmbloods, the appearance of an Oldenburg can vary depending on its exact parentage, but in general they are fitting sport horses, with long legs and a characterful head on a fairly long neck. Particular types are bred for dressage, jumping, or high-performance abilities, while others are aimed at the needs of amateurs. Oldenburgs can always be recognized by the "O" and crown brand on their left hip, with two identifying digits below. The usual Oldenburg coat colors are gray, black, brown, bay, and chestnut.

Scientific name	*Equus caballus*
Family	Equidae
Size	16–17.2 hands high; 1.62–1.75m (5.3–5.9ft) tall
Distribution	Originated in Germany
Habitat	Domesticated
Diet	Forage from hay or pasture, plus some grain
Breeding	Usually one foal

Nonius Horse

This breed is named after its foundation sire, a stallion called Nonius who was born in Calvados, Normandy, in 1810. Nonius had a Thoroughbred father, while his mother was a heavy Norman mare. Caught up in the Napoleonic wars, Nonius was captured and taken to the imperial stud farm at Mezohegyes in Hungary, where he began the Nonius breed. Nonius horses were originally intended as light military draft horses, but today the breed is used in leisure riding, driving sports, and limited agricultural work. The Nonius is usually too slow and heavy to compete at show jumping and dressage. Known for its even temperament and willingness to work, the Nonius displays a heavy head, powerful neck, broad back, and large joints. The most common coat colors are black, bay, and brown.

Scientific name	*Equus caballus*
Family	Equidae
Size	15.1–16.1 hands high; 1.53–1.63m (5–5.4ft) tall
Distribution	Originated in Hungary
Habitat	Domesticated
Diet	Forage from hay or pasture, plus some grain
Breeding	Usually one foal

Murgese Horse

With a history reaching back at least 500 years, the Murgese was developed from Arabian, native, and draft horses in the Apulia region of southeastern Italy. By the early twentieth century, the breed had nearly died out and was reestablished from a limited stock, leaving the modern horse probably more refined than the original Murgese. The horse frequently lives and breeds in a semiwild state in the region, foraging for food outdoors year-round, which has made the Murgese extremely tough and strong. The horse has a plain, light head; a sturdy neck; and a low-set tail. Its back is sometimes hollow and its legs are powerful and large-jointed. In Italy, the Murgese is used by the police force and is commonly kept for cross-country riding and light farmwork.

Scientific name	*Equus caballus*
Family	Equidae
Size	14–15 hands high; 1.42–1.52m (4.7–5ft) tall
Distribution	Originated in Italy
Habitat	Domesticated; rocky pastures
Diet	Forage from hay or pasture, plus some grain
Breeding	Usually one foal

Salerno Horse

The Salerno hails from the Italian province of Salerno, close to Naples, where it was developed in the late eighteenth century. The breed was championed by Charles III, king of Spain, Naples, and Sicily. The Salerno's key ancestors are the Andalusian and the now-extinct Neapolitan, making it a classic riding mount and fitting the breed to its use as a cavalry horse. In the twentieth century, Hackney and Thoroughbred blood were introduced, increasing the breed's size and refining it. Salernos have since competed internationally at show jumping and make good sporting horses. Salerno horses have a large head, a thick mane and tail, and a long, muscular neck. Their backs are long, their hindquarters are muscular, and they have slender but powerful legs. Salernos are usually found with chestnut, bay, or black coats.

Scientific name	*Equus caballus*
Family	Equidae
Size	16–17 hands high; 1.62–1.73m (5.3–5.7ft) tall
Distribution	Originated in Italy
Habitat	Domesticated
Diet	Forage from hay or pasture, plus some grain
Breeding	Usually one foal

San Fratello Horse

Also known as the Sanfratellano, this breed lives wild in the Nebrodi Mountains of Sicily. This rocky region, which has wide variations in temperature, has greatly influenced the horse, leaving it hardy, strong, and with great endurance. The San Fratello is always bay or black and is muscular and powerfully built. The breed is used locally as a pack and light draft horse, as well as a leisure mount. Very little is known about the ancestry of the horse. Local tradition holds that the San Fratello is descended from the ancient Sicilian horse, which was known through the Mediterranean and much admired by the ancient Greeks. In recent decades, Anglo-Arabian, Salerno, Nonius, and Thoroughbred blood has been introduced, but the horse's appearance remains much as it has for generations.

Scientific name	*Equus caballus*
Family	Equidae
Size	15–16 hands high; 1.52–1.62m (5–5.3ft) tall
Distribution	Originated in Sicily, Italy
Habitat	Rocky, mountainous pastures
Diet	Forage from pasture, plus some grain
Breeding	Usually one foal

Haflinger Horse

Also known as the Avelignese, this small horse was formally established in the village of Hafling, then in Austria, during the late nineteenth century. The village is today called Aveligna and lies in northern Italy. The horse gets its distinctive appearance from the crossing of Tyrolean ponies with Arabians and other, European breeds. Haflingers are chestnut, ranging from golden to liver, with a white or flaxen mane and tail. The breed has a delicate head, deep chest, muscular back, and strong legs. Haflingers from Austria and Italy are branded with an edelweiss design. The Haflinger is a popular horse for children but is also strong enough to make a good leisure horse for adults. It is used worldwide for harness and under-saddle events, including endurance riding, dressage, and show jumping.

Scientific name	*Equus caballus*
Family	Equidae
Size	13.2–15 hands high; 1.34–1.52m (4.5–5ft) tall
Distribution	Originated in Austria and northern Italy
Habitat	Domesticated
Diet	Forage from hay or pasture, plus some grain
Breeding	Usually one foal

Friesian Horse

Although the Friesian resembles a light draft horse, its agility and trainability make it increasingly popular for pleasure riding and even dressage, as well as driving. The breed displays a solid black coat color, occasionally with a small white star on the forehead. The thick, long mane and tail are accompanied by a feathering of silky hair on the lower legs. With its powerful but elegant build, compact body, and relatively short limbs, the Friesian carries itself with confidence. This dramatic appearance has allowed the breed to have starring roles in Hollywood action films such as *The Mask of Zorro* (1998). Originating in Friesland in the Netherlands, the Friesian is said to be descended from ancient forest-dwelling horses and may be among the ancestors of the Shire and Clydesdale.

Scientific name	*Equus caballus*
Family	Equidae
Size	14.2–17 hands high; 1.44–1.73m (4.9–5.7ft) tall
Distribution	Netherlands
Habitat	Domesticated
Diet	Forage from hay or pasture, plus some grain
Breeding	Usually one foal

Lusitano

Descended from the world-famous Andalusian horse, the Lusitano originated in Portugal and is named after Lusitania, the ancient Roman title for the region. The Andalusian and Lusitano were registered together in the Spanish stud book until the 1960s, when breeders of the Lusitano decided to emphasize its distinctive qualities. The Lusitano is the breed used in the first portion of a Portuguese bullfight, the *cavaleiro,* during which a horseman fights the bull while riding a specially trained horse. This has led to the Lusitano earning a reputation for bravery and calmness, as well as for bonding deeply with its rider. Lusitanos also compete internationally at dressage and show jumping. The Lusitano has powerful hindquarters, a broad chest, strong legs, and a low-set tail. A variety of solid colors are accepted, with many Lusitanos turning gray as they age.

Scientific name	*Equus caballus*
Family	Equidae
Size	15–16 hands high; 1.52–1.62m (5–5.3ft) tall
Distribution	Originated in Portugal
Habitat	Domesticated
Diet	Forage from hay or pasture, plus some grain
Breeding	Usually one foal

Andalusian Horse

The Andalusian is probably the most famous of all the Iberian horses. These breeds developed in Spain and Portugal over many millennia and have primitive indigenous horses such as the Sorraia among their ancestors. With its high, elegant stepping action, the Andalusian is particularly renowned for its skill at haute école dressage. In its native Spain, the Andalusian is also used in bullfighting, while this dashing breed makes many starring appearances in Hollywood films, including the *Lord of the Rings* trilogy. The Andalusian's most characteristic features are its compact build, lean head, large eyes, long neck, and flowing mane and tail. About 80 per cent of the breed are gray, while other accepted purebred colors are bay, chestnut, and black. The Andalusian is known for being sensitive, proud, and quick to learn.

Scientific name	*Equus caballus*
Family	Equidae
Size	15.2–16.2 hands high; 1.54–1.64m (5.1–5.5ft) tall
Distribution	Originated in Spain
Habitat	Domesticated
Diet	Forage from pasture, plus some grain
Breeding	Usually one foal

Freiberger Horse

Also known as the Franches Montagnes, the Freiberger is a versatile light draft horse that is also used for pleasure and competitive riding. The breed was developed in Switzerland by crossing draft horses such as the Bernese Jura and Ardennais with Thoroughbred and Arabian hotbloods. The result is a horse with a fairly heavy but small head, an arched and muscular neck, sloping shoulders, and a strong, straight back. Modern breeding has largely taken away the feathering around the fetlocks, and has moved toward a lighter and more refined Freiberger that is favored as a leisure mount. A more muscular type is still used by upland farmers in the Jura region and by the Swiss army. The usual Freiberger coat colors are bay or chestnut. The breed is known for being active, calm, and sure-footed.

Scientific name	*Equus caballus*
Family	Equidae
Size	14.3–15.3 hands high; 1.45–1.55m (4.9–5.3ft) tall
Distribution	Originated in Switzerland
Habitat	Domesticated
Diet	Forage from pasture, plus some grain
Breeding	Usually one foal

Clydesdale Horse

This rare eighteenth-century breed of draft horse originated in Clydesdale, Scotland. A Clydesdale has a large head with a convex, or Roman, nose. Its chest is deep, the rump well muscled, and the legs long and strong. The most characteristic features are the very wide hoofs, covered with an abundance of feather—the long hair that falls from below the knees. Clydesdales may be of many colors, including chestnut, bay, and black, always with white feet and a large white blaze on the face. Aside from their use in draft horse showing, in which horses are judged on their performance in harness and halter, Clydesdales are increasingly popular horses to be ridden under saddle.

Scientific name	*Equus caballus*
Family	Equidae
Size	16–18 hands high; 1.63–1.83m (5.3–6ft) tall
Distribution	Originated in Scotland
Habitat	Domesticated
Diet	Forage from hay or pasture, plus some grain
Breeding	Usually one foal

Hackney Horse and Pony

The Hackney has been developed for carriage driving and is known for its ability to trot at a fast speed for long periods of time, but is also a well-liked breed for riding, show jumping, and dressage. The breed is recognized in both horse and pony sizes, with the latter being a popular show pony, both in harness and under saddle. Hackney ponies are known for being healthy and relatively easy to keep. Their good temper and enjoyment of gentle interaction makes them a popular choice for children. Most recognizable when in motion, the Hackney is known for its showy, high-kneed movement. The horse has a handsome head and large, expressive eyes. The tail is set high and the legs are strong. Hackneys may be of any solid color, often with white markings.

Scientific name	*Equus caballus*
Family	Equidae
Size	Horse: 14.2–16.2 hands high; 1.44–1.54m (4.8–5.5ft) tall Pony: 12–14 hands high; 1.21–1.42m (4–4.7ft) tall
Distribution	Originated in England
Habitat	Domesticated
Diet	Forage from hay or pasture, plus some grain
Breeding	Usually one foal

Shire Horse

The tallest of the draft horse breeds, the Shire horse is said to be descended from the great war stallions brought to England by William the Conqueror in 1066. A Shire horse holds the record for being the world's tallest horse: in the 1850s, an English horse named Sampson was measured at more than 21.2 hands high (2.15 metres; 7.2 feet). Today, the breed competes in draft horse shows worldwide and is also used by a few English brewers to pull beer carts. The Shire horse is powerful and muscular with a broad back and long legs. Its girth can be as large as 8 feet. Colors are usually gray, bay, and black. Bay and black horses should have feathered white stockings. The Shire horse is known for its docile brown eyes.

Scientific name	*Equus caballus*
Family	Equidae
Size	17–19 hands high; 1.73–1.93m (5.7–6.3ft) tall
Distribution	Originated in England
Habitat	Domesticated
Diet	Forage from hay or pasture, plus some grain
Breeding	Usually one foal

Thoroughbred Horse

Although the term "thoroughbred" is sometimes used to refer to any purebred horse, the Thoroughbred is, in fact, a renowned breed known for its use in racing. The Thoroughbred originates in seventeenth-century England, where it was begun by crossing native mares with Arabian stallions. Thoroughbreds always have single-color coats, found in gray, black, bay, brown, chestnut, roan, palomino, and—very rarely—white. The horses have long legs, lean bodies, deep chests, and finely boned heads on long necks. The Thoroughbred is known as a hot-blooded breed, developed for its agility and speed while displaying a spirited temperament. In addition to being racehorses, Thoroughbreds make fine family riding horses, show jumpers, dressage horses, combined training horses, and polo ponies. Thoroughbreds are known for having a high rate of health problems, while competing horses are injury-prone.

Scientific name	*Equus caballus*
Family	Equidae
Size	15.2–17 hands high; 1.54–1.73m (5.2–5.6ft) tall
Distribution	Originated in England
Habitat	Domesticated
Diet	Forage from hay or pasture, plus some grain
Breeding	Usually one foal

Akhal-Teke Horse

The Akhal-Teke is one of the oldest surviving horse breeds, originating in Turkmenistan. This elegant horse is known as a hotblood, which means it was bred for agility and speed. The breed is also highly resilient because it is adapted to the harsh climate and conditions of the central Asian desert, where food and water can often be scarce. These characteristics make the Akhal-Teke a popular choice for endurance riding, eventing, racing, show jumping, and dressage. The breed is famous for those individuals who have a pale, golden buckskin color with a strong metallic sheen. This patterning is thought to have been useful camouflage in the desert. Other colors include bay and palomino, which also tend to shimmer, as well as black, chestnut, and gray. The Akhal-Teke has a fine head, long ears, and almond-shaped eyes.

Scientific name	*Equus caballus*
Family	Equidae
Size	14.3–16.3 hands high; 1.45–1.66m (5–5.5ft) tall
Distribution	Originated in Turkmenistan; found in the United States, Russia, and Germany
Habitat	Domesticated; adapted to desert conditions
Diet	Forage from hay or pasture, plus some grain
Breeding	Usually one foal

Appaloosa Horse

Known for its spotted coat, the Appaloosa was developed by the Nez Perce people of what is now the northwestern United States. Horses reached the region in about 1730, when the Nez Perce began to turn breeding into an art form, coming to emphasize spotting in their program. In 1877 the Nez Perce and their horses were decimated by the U.S. Army in the Nez Perce War. The surviving stock was revitalized in the twentieth century. As spotting is the preferred identifying factor, there are several body styles found in the Appaloosa, with different types being more suited to ranch work or English disciplines. There are thirteen recognized base coat colors, which are overlaid by various spotting patterns, from leopard (dark spots on a white body) to snowflake (white spots on a dark body).

Scientific name	*Equus caballus*
Family	Equidae
Size	14.2–16 hands high; 1.44–1.63m (4.8–5.3ft) tall
Distribution	Originated in the United States
Habitat	Domesticated
Diet	Forage from hay or pasture, plus some grain
Breeding	Usually one foal

Morgan Horse

All Morgan horses are descendants of a stallion named Figure, born in Massachusetts in 1789 and at one time owned by a man named Justin Morgan. Figure himself may have been descended from Thoroughbred and Friesian horses, but his exact ancestry is unknown. Figure was known for his impressive strength, build, and athleticism as well as his good temperament. His stud services were in exhausting demand in New England for many years. Today the Morgan horse has a compact build with strong limbs and an expressive face. A wide range of colors are found, with the most common being bay, chestnut, and black. The Morgan is extremely versatile and competes at high levels in western events, dressage, show jumping, eventing, and driving. The Morgan is the state horse of both Vermont and Massachusetts.

Scientific name	*Equus caballus*
Family	Equidae
Size	14.1–15.2 hands high; 1.43–1.54m (4.8–5.2ft) tall
Distribution	Originated in the United States
Habitat	Domesticated
Diet	Forage from hay or pasture, plus some grain
Breeding	Usually one foal

American Quarter Horse

The most popular horse in the United States, the quarter horse is named for its ability to outdistance other breeds at races of a quarter of a mile or less. The breed is descended from the Thoroughbreds taken to America by English colonists and then crossbred with the horses that they found there, such as the Chickasaw, which was itself descended from the Iberian horses of Spanish colonists. As American settlement moved westward, the quarter horse was crossed with the feral horses known as mustangs, and developed into a great ranch horse. The quarter horse is today widely used in western riding events, particularly those involving cattle, as well as in English disciplines such as dressage. The breed displays a small, refined head; a well-muscled body; and powerful hindquarters.

Scientific name	*Equus caballus*
Family	Equidae
Size	14–16 hands high; 1.42–1.63m (4.7–5.3ft) tall
Distribution	Originated in the United States
Habitat	Domesticated
Diet	Forage from hay or pasture, plus some grain
Breeding	Usually one foal

Standardbred Horse

Best known for its excellence in harness at a trot or a pace, the Standardbred was named for its ability to trot a mile under the standard of two minutes and thirty seconds. Today's Standardbreds can race much faster, pacing the mile in less than one minute and fifty seconds. There are two types of Standardbreds: trotters and pacers. The trotter's racing gait is the trot, in which its legs move in diagonal pairs, left fore with right hind. The pace is a lateral gait, left fore with left hind. The breed also performs all other gaits, and pacers can be trained to trot. Standardbreds are used worldwide for harness racing as well as for leisure riding and sporting. This well-muscled breed is considered to be placid and easy to train.

Scientific name	*Equus caballus*
Family	Equidae
Size	14.1–17 hands high; 1.43–1.73m (4.8–5.7ft) tall
Distribution	Originated in the United States
Habitat	Domesticated
Diet	Forage from pasture, plus some grain
Breeding	Usually one foal

Criollo Horse

The Criollo has its roots in a shipment of Andalusian horses that arrived in the newly founded Buenos Aires, Argentina, in 1536. In 1541, indigenous peoples forced the Spaniards to abandon the settlement, and the Andalusians were set loose. When Buenos Aires was reestablished in 1580, these feral horses had developed into a hardy breed that was able to survive extreme weather with little water and just sparse, dry grass. These characteristics remain intact to this day. Over the centuries, the Criollo was crossbred with various breeds, including the Thoroughbred and Chilean. The modern Criollo is strong-bodied with short legs and hard feet. Common coat colors are dun, bay, brown, black, chestnut, and buckskin. The breed is widely used in South America as a ranch horse, but also makes a popular leisure and endurance horse.

Scientific name	*Equus caballus*
Family	Equidae
Size	13.1–14.3 hands high; 1.33–1.45m (4.4–4.9ft) tall
Distribution	Originated in Argentina
Habitat	Domesticated
Diet	Forage from pasture, plus some grain
Breeding	Usually one foal

Peruvian Paso Horse

Bred for its smooth gait, the Peruvian Paso is ideal for leisure, endurance, and trail riding. Horses arrived in Peru during the Spanish conquest and, since there was no widespread livestock-based economy, were used mainly for transportation over long distances. Peruvian breeders came to breed selectively for gait, hardiness, and temperament. In place of a bouncing trot, the Peruvian Paso has an ambling four-beat gait that is between a walk and a canter. It is performed laterally, with left hind, left fore, right hind, right fore. Physically, the breed is elegant yet powerful, with a heavy neck and a low-set tail. The mane and forelock are particularly abundant. Perhaps more important than all of these characteristics, breeders consider that a key trait of the horse is its brio—its spirit, pride, and willingness to please.

Scientific name	*Equus caballus*
Family	Equidae
Size	14.1–15.2 hands high; 1.43–1.54m (4.8–5.2ft) tall
Distribution	Originated in Peru
Habitat	Domesticated
Diet	Forage from hay or pasture, plus some grain
Breeding	Usually one foal

Anglo-Arabian Horse

As its name suggests, the Anglo-Arabian is the result of crossbreeding between the English Thoroughbred and the Arabian horse. To be classified as an Anglo-Arabian, the horse must have at least one-quarter, but no more than three-quarters, Arabian blood. Anglo-Arabians have variable size and appearance, depending on the degree to which they descend from one breed or the other. An Anglo-Arabian is usually taller than the average Arabian and has slightly less physical refinement. The largest individuals are the result of a pairing between a Thoroughbred mare and an Arabian stallion. The finest examples of the breed display the physique and stamina of the Arabian alongside the speed of the Thoroughbred. The usual colors are chestnut, gray, or bay. The breed offers very popular riding, sporting, and eventing horses.

Scientific name	*Equus caballus*
Family	Equidae
Size	15.2–16.3 hands high; 1.54–1.66m (5.2–5.6ft) tall
Distribution	Worldwide
Habitat	Domesticated
Diet	Forage from hay or pasture, plus some grain
Breeding	Usually one foal

Dülmen Pony

The only surviving pony breed native to Germany, the Dülmen is believed to have developed from primitive native ponies. The earliest mention of the Dülmen ponies, which are found near to the town of Dülmen in Westphalia, dates back to the early fourteenth century. The ponies lived in wild herds across Westphalia until the mid-nineteenth century, when much of the land was enclosed for farming. Today, only one semiwild herd, owned by the Duke of Croy, still roams the area. All the Dülmens ridden around the world originate from this herd. The breed remains unrefined in type, with a short neck and legs and a stocky build. The most common coat colors are dun, chestnut, brown, and black. The Dülmen is known for making a good children's pony and for doing well in harness.

Scientific name	*Equus caballus*
Family	Equidae
Size	12–13 hands high; 1.21–1.32m (4–4.3ft) tall
Distribution	Originated in Germany
Habitat	Semiwild; domesticated
Diet	Forage from hay or pasture, plus some grain
Breeding	Usually one foal

Dartmoor Pony

This kindly and gentle breed of pony is popular as a first horse for children, but is also capable of carrying adults. Dartmoors are commonly used for leisure riding, jumping, and driving. The Dartmoor has a small head, short legs, and a thick mane and tail. It is accepted in bay, chestnut, brown, black, gray, or roan. As well as being bred in Britain, North America, and Europe, feral Dartmoor ponies live in Dartmoor National Park in southwestern England. The ancestors of these ponies were turned loose after the mechanization of the local mining industry in the mid-eighteenth century. The harsh climate on the moor has contributed to the breed's stamina and hardiness. Only an estimated 5,000 Dartmoors survive today, leading to the pony's classification as a rare breed.

Scientific name	*Equus caballus*
Family	Equidae
Size	11.1–12.2 hands high; 1.13–1.24m (3.7–4.2ft) tall
Distribution	Originated in the United Kingdom
Habitat	Moorland; domesticated
Diet	Forage from hay or pasture, plus some grain
Breeding	Usually one foal

Fell Pony

Originating on the fells of the Lake District of northeastern England, the Fell pony makes an excellent, versatile family pony, suitable for both adults and children. The breed may have its roots in the crossing of Roman war stallions with local Celtic ponies during the Roman occupation of Britain. With their muscular bodies, strong legs, and easygoing temperament, Fell ponies have been used as pack horses right up to modern times. The breed is known for being sure-footed on uneven ground and providing a comfortable ride. In the United Kingdom, the Riding for the Disabled movement commonly uses Fell ponies. These ponies are also widely used for trekking and driving, as well as for jumping to a local show standard. Fell ponies are usually seen in black, gray, brown, and bay.

Scientific name	*Equus caballus*
Family	Equidae
Size	13–14 hands high; 1.32–1.42m (4.3–4.7ft) tall
Distribution	Originated in the United Kingdom
Habitat	Domesticated
Diet	Forage from hay or pasture, plus some grain
Breeding	Usually one foal

Shetland Pony

Shetland ponies are a naturally occurring breed from the wind-lashed Shetland Isles, which lie northeast of mainland Scotland. Used for centuries in agriculture, Shetlands are hardy and extremely strong for their size. In the mid-nineteenth century, as the demand for coal increased, Shetlands were imported onto mainland Britain to work as pit ponies. The breed displays a small head with alert ears, a short and muscular neck, a stocky body, and sturdy legs. Shetlands were first imported to North America in 1885 and were later bred with pony breeds such as the Hackney and Welsh. The American Shetland is more refined than the traditional Shetland, with longer legs and a narrower back. With their gentle nature, Shetlands make very popular mounts for children and are also used for pleasure driving.

Scientific name	*Equus caballus*
Family	Equidae
Size	7–11.2 hands high; 0.71–1.14m (2.3–3.8ft) tall
Distribution	Originated in the United Kingdom
Habitat	Domesticated
Diet	Forage from hay or pasture, plus some grain
Breeding	Usually one foal

Baudet de Poitou Donkey

This French breed of donkey is one of the world's rarest: there are perhaps fewer than 180 purebreds in existence. The Poitou is instantly recognizable for its shaggy coat, known as a *cadanette*, which hangs in matted tangles when ungroomed. The Poitou's coat is dark brown or black, with a white underbelly, white nose, and rings around its eyes. The Poitou was developed for its large size, in order to breed bigger and stronger working mules—the offspring of a male donkey and a female horse, which are themselves sterile. Mules are deemed to have the endurance of a donkey and the vigor of a horse. A Poitou's ears are long, its head is large, and its feet are big. Today, French and U.S. breeding programs enable private owners to keep a Poitou.

Scientific name	*Equus asinus*
Family	Equidae
Size	15–16 hands high; 1.52–1.63m (4.4–5ft) tall
Distribution	Originated in France
Habitat	Domesticated
Diet	Grass hay, plus carrots as treats and a mineral salt block to lick
Breeding	Usually one foal

Oriental Fire-Bellied Toad

The Oriental toad's eponymous reddish-orange belly is mottled with dark brown to black. The upper side of its body is usually bright green with black mottling. In the wild, the Oriental toad's diet includes aquatic arthropods containing carotene, which keeps its belly brightly colored. In captivity, feeding carrots to the toad's prey insects, such as crickets, will help to maintain its coloration. The bright color warns predators that this toad is toxic: if threatened, it secretes toxin through the skin of its belly. For this reason, this amphibian should not be kept with most other types of toads, and owners should wash their hands after handling. Commonly captive-bred for the pet trade, this hardy amphibian is likely to do well in a semiaquatic terrarium or an aquarium with several floating islands.

Scientific name	*Bombina orientalis*
Family	Bombinatoridae
Size	3.8–5cm (1.5–2in) long
Distribution	Korea and northeastern China
Habitat	Shallow pools in warm and humid forests
Diet	Small crickets, grubs, and worms
Breeding	40–100 eggs laid in aquatic plants

European Green Toad

As European green toads are relatively easy to care for, they have become popular pets. They are pale gray with olive or dark green patches interspersed with red dots. This species can change color very quickly and markedly in response to its environment. Like most toads, these amphibians secrete a toxin in glands on the back of their neck, so owners should wash their hands after handling them. In fact, these toads have sensitive skin, so they should not be handled often. European green toads should be housed in at least a 60-litre (13-gallon) planted enclosure kept at 27°C (80°F) during the day and about 5°C (10°F) cooler at night. The humidity level should be close to 50 per cent, and a water receptacle large enough for the toad to soak in needs to be provided.

Scientific name	*Bufo viridis*
Family	Bufonidae
Size	7.5–15cm (3–6in) long
Distribution	Southern and central Europe, Mediterranean islands, North Africa, and central Asia
Habitat	Range of habitats from semidesert to mountains
Diet	Crickets, wax worms, earthworms, and red worms
Breeding	9,000–15,000 eggs at a time; survival rate of perhaps 1 in 400

Blue Poison Dart Frog

This stunning frog is colored bright blue to warn would-be predators that it is poisonous. Its skin toxins are not actually produced by the frog itself: they are taken from its insect prey in the wild and deposited in the skin. For this reason, frogs of this species that are raised in captivity are not poisonous. Unlike most frogs, these amphibians lay their eggs on land, usually under a rock or log near a stream. Blue poison dart frogs are fairly easy to keep, with some surviving for up to eight years, and are highly recommended for beginners. As many as four of these frogs will be comfortable in a heavily planted 110-litre (24-gallon) terrarium. The terrarium should be humid and kept at 24–27°C (75–80°F) and contain a shallow dish of clean water.

Scientific name	*Dendrobates azureus*
Family	Dendrobatidae
Size	2.5–6cm (1–2.5in) long
Distribution	South America: Suriname
Habitat	Rain forest, under mossy rocks near streams
Diet	Tiny insects: pinhead crickets, fruit flies, rice flour beetle larvae, and termites
Breeding	5–10 eggs at a time

Barking Tree Frog

The largest tree frog native to the United States, this amphibian is known for its loud barking call. Males have an enlarged vocal sac. Individuals may be green, brown, yellow, or gray but can be recognized from the darker round spots and irregular lateral stripes on their bodies. These frogs' colors and patterns can change quickly depending on their environment. In the wild, the barking tree frog breeds in shallow pools and may burrow in sand in hot weather. It is less active during the day, when it may climb high up into trees using its prominent toe pads. Several of these tree frogs can be kept in a damp terrarium with a capacity of 60–90 litres (13–20 gallons), with a shallow bowl of water for bathing. If kept clean and healthy, the barking tree frog can live for ten years in captivity.

Scientific name	*Hyla gratiosa*
Family	Hylidae
Size	5–6cm (2–2.5in) long
Distribution	United States: South Carolina to Florida and Louisiana
Habitat	Coastal areas
Diet	A variety of insects
Breeding	2,000 eggs at a time; survival rate of perhaps 1 in 400

Clown Tree Frog

These beautiful and tiny tree frogs are generally a hardy and easy-to-keep species. They have a dark and slender body dabbed with irregular blobs of yellow or cream. The undersides of their legs are orange, as is the webbing between their toes. Several of these tree frogs can be housed together in a damp, planted terrarium complete with diagonal and horizontal perches. These frogs are naturally shy and inactive during the day. They tend to find a hiding spot and sleep the day away, and only begin to get active at dusk. Some owners report that a good way to watch their activity is to create a routine that they come to expect: turn down the lights at dusk, give them a misting, and offer food—then watch them come out of hiding.

Scientific name	*Hyla leucophyllata*
Family	Hylidae
Size	2.5cm (1in) long
Distribution	Amazon basin
Habitat	Lowland rain forest
Diet	Moths and tiny insects
Breeding	600 eggs at a time; survival rate of perhaps 1 in 400

Asian Painted Frog

This bullfrog is known by many names, including the chubby frog, bubble frog, banded bullfrog, and rice frog. It belongs to the narrow-mouthed frog group and has a rich purple-brown back, distinctive copper-colored stripes down its sides, and a cream stomach. It is known as a bullfrog for its loud, mooing call. It is a ground dweller, and in the wild it will usually hide under leaf litter during the day, waiting until the evening to search for food. Like most other narrow-mouthed frogs, the Asian painted frog can expand itself when threatened and defensively secrete toxic substances. Several of these frogs can be kept together in a large, damp terrarium with a substrate of peaty soil or potting soil with sphagnum moss. The tank must be covered and kept at a temperature of 27–29°C (80–85°F).

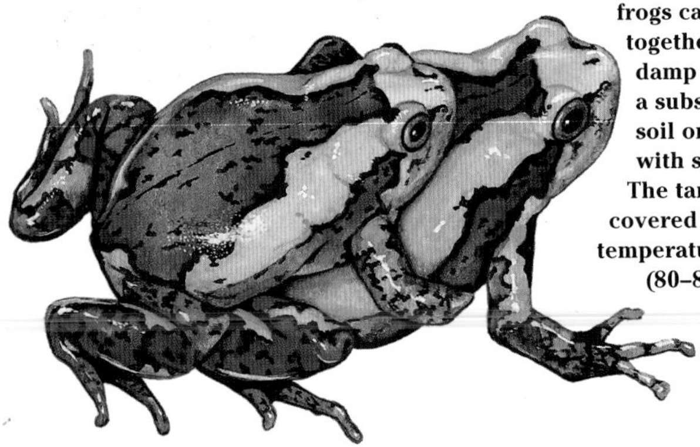

Scientific name	*Kaloula pulchra*
Family	Microhylidae
Size	5–7.5cm (2–3in) long
Distribution	Southeast Asia: Malaysia and Indonesia to China and Taiwan
Habitat	Forest floor or rice fields to inside homes
Diet	Tiny insects, including pinhead crickets, aphids, and termites
Breeding	1,000 eggs at a time; survival rate of perhaps 1 in 400

Giant African Bullfrog

This frog is well known for its immense size, with a male weighing up to 2 kilogrammes (4.5 pounds). Its skin is olive green with bumpy skin folds along the back. Males have a yellow throat and are about twice the size of females. This frog is known for being quite aggressive and can give a very painful bite. In the wild, African bullfrogs dig a hole in the ground during the dry season, where they build a cocoon to prevent evaporation of their bodily fluids. Once the rainy season starts, the cocoon splits open and the frog makes its way to a pool of water to breed. This is when the males make their throaty, grunting calls. In the home, similarly sized adults can be kept together, provided that the tank is large enough for them.

Scientific name	*Pyxicephalus adspersus*
Family	Ranidae
Size	Males: 18–24cm (7–9.5in) long; females: 7.5–11cm (3–4.5in) long
Distribution	Central and southern Africa
Habitat	Grassland
Diet	Insects, fish, and only an occasional mouse
Breeding	3,000–4,000 eggs at a time; survival rate of perhaps 1 in 400

American Bullfrog

The largest of the North American true frog family, the American bullfrog is an aquatic species. These frogs are brownish-green to green and have fully webbed hind feet. Males have a visible eardrum larger than their eyes, while in females the eardrum is the same size as the eyes. Fertilization is external in true frogs. Male American bullfrogs take a position in their territory, attracting females with their loud call, similar to the roar of a bull. The male then rides the female, simultaneously releasing his sperm as she deposits her eggs in the water. These frogs are cannibalistic, so frogs of different sizes should never be housed together. As these are active frogs, a pair of adults will need a tank of 150–220 litres (34–48 gallons).

Scientific name	*Rana catesbeiana*
Family	Ranidae
Size	10–20cm (4–8in) long
Distribution	North America
Habitat	Shorelines of permanent bodies of water
Diet	Insects, worms, fish, and an occasional mouse
Breeding	20,000 eggs at a time; survival rate of perhaps 1 in 400

Pickerel Frog

The pickerel frog is a member of the North American leopard frog group, distinguished by their green skins and prominent black spots. The other frogs within this group have circular spots, but the pickerel has rectangular spots, which may sometimes merge to form a long rectangle on the back. Another distinguishing feature is the orange-yellow flash pattern on the inner skin of the hind legs. These frogs are semiaquatic and need both water deep enough to submerge in and an area of dry ground in their tank. Pickerels are genetically programmed to hibernate, so will slow down and may stop eating for about three months in the winter. Pickerel frogs should never be housed with other species because their skin secretions are toxic. Owners are also advised to wash their own hands after handling.

Scientific name	*Rana palustris*
Family	Ranidae
Size	7.5–10cm (3–4in) long
Distribution	Central and eastern United States; southeastern Canada
Habitat	Close to permanent bodies of water
Diet	Crickets, wax worms, fly larvae, and earthworms
Breeding	3,000 eggs at a time; survival rate of perhaps 1 in 400

Marsh Frog

The largest European frog, the marsh frog, also called the laughing frog, is mainly a water dweller. It varies from dark green to brown-gray, but a paler green line along the center of the back is usually present. Males have gray vocal sacs. Western European marsh frogs are usually dark green with darker spots. In the wild, marsh frogs eat dragonflies, spiders, earthworms, slugs, mice, and even salamanders and fish. Only males make a noise, with their call being a loud, ducklike croak. As many as four adults can be housed in a semiaquatic 110-litre (24-gallon) tank. A good method of setting this up is to make a large floating island for the frogs to comfortably climb onto. Numerous aquatic plants can be arranged to provide hiding places, in a base substrate of gravel.

Scientific name	*Rana ridibunda*
Family	Ranidae
Size	11–18cm (4.5–7in) long
Distribution	Southwestern and eastern Europe; Asia: Russia to Pakistan and western China
Habitat	Marshes, meadows, and woods close to still water
Diet	Crickets, mealworms, earthworms, wax worms, and curly flies
Breeding	1,000–4,000 eggs; survival rate of perhaps 1 in 400

European Common Frog

This frog is common across Europe, as far north as the Arctic Circle and as far south as Italy. The common frog's dark-blotched back ranges from olive green through brown to gray or yellow, with some color change depending on its surroundings. The underbelly is white or yellow, with females often sporting a more orangey shade. During the mating season, males are generally lighter or gray in color, frequently with a bluish throat. Males have swellings known as nuptial pads on their first fingers, used for gripping females during mating. In the wild, the common frog hibernates during the cold winter months, in the mud at the bottom of ponds, in burrows, or in running water. In a semiaquatic home aquarium, this species is likely to stay active all year.

Scientific name	*Rana temporaria*
Family	Ranidae
Size	5–9cm (2–3.5in) long
Distribution	Europe: Southern Italy to the Arctic Circle; Ireland to the Urals
Habitat	Close to freshwater
Diet	Crickets, fruit flies, houseflies, worms, snails, and slugs
Breeding	Up to 4,000 eggs; survival rate of perhaps 1 in 400

Fire Salamander

There are fifteen recognized subspecies of these striking salamanders, most being black with spots or stripes of bright yellow. In some subspecies, red and orange may replace or mix with the yellow. Adults are able to extrude toxic skin secretions—gloves and goggles should be worn the first few times a salamander is held. Most amphibians lay eggs in water, but almost all of the fire salamander subspecies are ovoviviparous, which means that embryos develop within eggs that remain inside the mother's body until they hatch. The yellow-striped fire salamander gives birth to live young. The fire salamander is primarily a land dweller, although its larvae are dependent on freshwater to develop. Several fire salamanders can be kept together in a cool, semiaquatic 80-litre (18-gallon) aquarium. Their average lifespan is ten to twenty years.

Scientific name	*Salamandra salamandra*
Family	Salamandridae
Size	11–20cm (4.5–8in) long
Distribution	Southern and central Europe
Habitat	Hilly woodland close to brooks
Diet	Insects, worms, and lean raw meat
Breeding	20–70 live larvae

Crested Newt

The four species of crested newts are threatened in the wild and are protected by legislation, so captive-bred specimens should always be obtained. Captive-bred alpine crested newts (*Triturus carnifex*), northern crested newts (*T. cristatus*), and Balkan crested newts (*T. karelinii*) are commonly available, with Danube crested newts (*T. dobrogicus*) appearing more rarely. These newts earn their name from the crest that appears along the male's back during the breeding season. Once the season is over, the crest is absorbed into the body. Newts start their lives in water before moving onto land at around four months old, after they lose their gills and have grown legs. Newts return to the water to mate and hunt for food. In captivity, crested newts can do well in a spacious semiaquatic terrarium.

Scientific name	*Triturus genus*
Family	Salamandridae
Size	13–18cm (5–7in) long
Distribution	Europe and western Asia, depending on species
Habitat	Still or slow-moving water in scrub, grassland, or woodland
Diet	Earthworms, wax worms, and white worms
Breeding	200–600 eggs laid in aquatic vegetation

European Honeybee

The European, or western, honeybee is the species usually kept by beekeepers, with the Italian subspecies being the most commonly kept bee in temperate climates. Italian bees are considered to be gentle and not highly likely to swarm. Due to centuries of selective breeding, they produce a large surplus of honey for the beekeeper to collect. The summer population of a thriving hive may be 40,000–80,000 bees. Italian bee colonies tend to maintain large winter populations, and so require more winter feeding than other temperate subspecies. Beekeepers commonly supply a hive for the colony to live in, with some hive types requiring a permit. Expert knowledge of bees and their habits is the best defense against stings, but most beekeepers also wear some form of protective clothing.

Scientific name	*Apis mellifera*
Family	Apidae
Size	1.3cm (0.5in) long
Distribution	Kept commercially worldwide
Habitat	Movable framed hive
Diet	Sugar syrup needed at certain times of year
Breeding	Thousands of eggs laid by the queen bee

Madagascar Hissing Cockroach

One of the largest cockroaches in the world, these insects make popular pets, as they are calm, not known for biting, and create no unpleasant odor with their feces. Hissing cockroaches are known for their unique ability to hiss by forcing air through the breathing pores on their abdomen. They will hiss when disturbed, and males will also hiss when challenging each other. Unlike most cockroaches, hissers are wingless but are good climbers and can even scale the glass sides of a terrarium. A roomy, lidded glass tank can be provided for a group of hissing cockroaches, with a substrate of newspaper or coconut fiber. Some people experience an allergic reaction to the odor of the hissing cockroach, so limited inhalation and washing one's hands after handling are recommended.

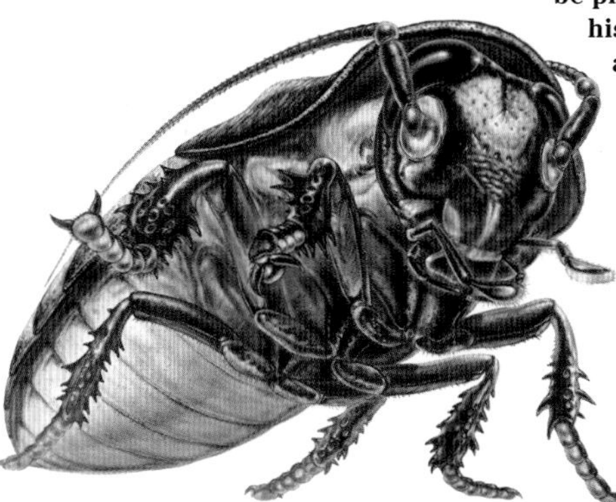

Scientific name	*Gromphadorhina portentosa*
Family	Blaberidae
Size	7–10cm (3–4in) long
Distribution	Madagascar; permits required for ownership in some states
Habitat	Rotting logs in the rain forest
Diet	Finely ground chick meal or dry dog food and fresh vegetables
Breeding	30–60 live young

Ladybird

There are more than 5,000 ladybird species, including the seven-spot ladybird (pictured). Ladybirds have oval bodies and are often orange or red with black spots, but numerous patterns occur. Children may be fascinated to observe the ladybird life cycle. The ladybird begins as an egg before developing into a larva. The larva will shed its skin up to seven times as it outgrows its exoskeleton (external skeleton). From the larval stage, a ladybird enters the pupal stage before finally emerging as an adult after about six weeks. Ladybirds are often sold for use in the garden, or can be caught entirely for free. A terrarium or bug box can be used as a home. Put moist foliage or a damp paper towel inside so that the ladybirds can drink.

Scientific name	*Coccinella* genus
Family	Coccinellidae
Size	0.25–1cm (0.1–0.4in) long
Distribution	Worldwide
Habitat	Gardens, forests, fields, and grasslands
Diet	Aphids, mealybugs, sugar water, and moistened raisins
Breeding	1,000–2,000 eggs

Leaf-Cutter Ant

There are fifteen species of leaf-cutter ants in the genus *Atta*, all found in the warmer regions of the Americas. These fascinating ants cultivate special structures called gongylidia, produced by a fungus that grows in the ants' nest. The ants feed the fungus on freshly cut plant material, collecting only those leaves on which the fungus thrives. A leaf-cutter colony can contain more than 8 million ants, which are divided into castes: minims, the smallest workers, which tend the brood and the fungus gardens; minors, slightly larger workers, which defend the foraging lines; mediae, the foragers; majors, the largest worker ants, which act as soldiers; and just one fertile queen. To keep a colony of leaf-cutters at home, plenty of room will be needed for a large nest tank and feeding tables.

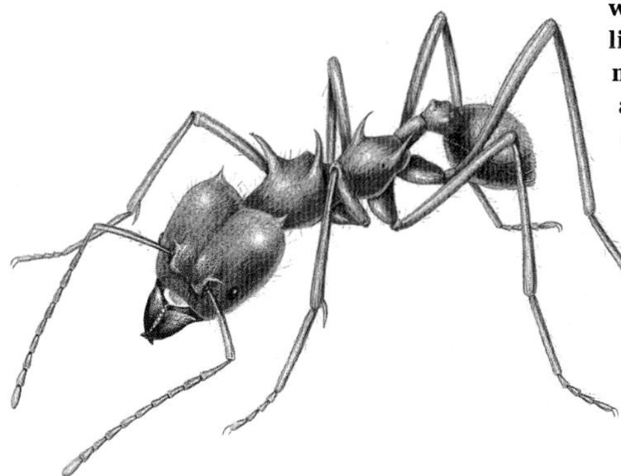

Scientific name	*Atta* genus
Family	Formicidae
Size	Worker: 0.1–0.8cm (0.05–0.3in) long; queen: up to 3.8cm (1.5in) long
Distribution	Central and South America: Mexico to Argentina
Habitat	Forest
Diet	Lepiotaceae fungi, fed on a variety of leaves
Breeding	Queen lays eggs throughout the year

Field Cricket

There are several species of field crickets in the genus *Gryllus*, including *Gryllus bimaculatus* and *Gryllus campestris* (pictured). The *Gryllus campestris* species is endangered in Britain and should not be taken from the wild there. *Gryllus bimaculatus* is often raised in captivity as a live food for exotic pets. Male field crickets are known for their mating song, or chirping, during the summer months. Females consider each song before plumping for the one they like best. When a male senses that a female is drawing near, he makes a gentler courting song. In captivity, field crickets should be kept in a terrarium planted with clumps of grass, frequently replaced as the grass is eaten. Drinking water can be provided in a shallow container such as an upturned lid.

Scientific name	*Gryllus* genus
Family	Gryllidae
Size	1.8–3cm (0.7–1.2in) long
Distribution	Worldwide
Habitat	Short, warm, tussocky grasslands
Diet	Grass, fruit, and fish food
Breeding	Eggs laid in soil in the fall

Stick Insects

Stick insects are highly popular with hobbyists, and more than 300 species are bred in captivity. Among the most commonly kept are the Indian stick insect (*Carausius morosus*) and northern stick insect (*Diapheromera femorata*). Indian stick insects in particular are considered hardy and easy to care for. Stick insects have developed camouflage to its furthest extent, mimicking the twigs of trees, both in form and coloration. A group of stick insects will live happily in a tall, wooded terrarium of at least 14 litres (3 gallons). Care depends on the species, but most species need plenty of humidity, so be prepared to spray the terrarium every day, and lay a few pieces of water-soaked paper towels on the tank floor. Many species have been known to reproduce by parthenogenesis, without mating.

Scientific name	*Carausius, Dares, Diapheromera, Phyllium,* and other genera
Family	Heteronemiidae, Phasmatidae, and others
Size	5–30cm (2–12in) long
Distribution	Temperate and tropical regions
Habitat	In vegetation
Diet	Variety of leafy branches, depending on species
Breeding	100–400 eggs laid on the ground

Stag Beetle

This beetle owes its name to the male's mandibles, which look like the antlers of a stag. Stag beetles are very popular among hobbyists, experienced and novice alike. Some species are endangered and should not be taken from the wild, including *Lucanus cervus* (pictured), which is native to Europe, and the highly prized king stag beetle, *Phalocragnathus muelleri*, of Australia. Luckily, king stag beetles are commonly bred in captivity. The majority of stag beetles are black or reddish-brown, but the king is a shimmering green. In areas where the stag beetle population is endangered due to loss of habitat, children can help by creating a backyard breeding site: bury a bucket perforated with 4-centimetre (1.5-inch) holes and filled with soil and wood chips, so that females can lay their eggs.

Scientific name	*Aesalinae, Lampriminae, Lucaninae,* and *Syndesinae* subfamilies
Family	Lucanidae
Size	2.5–7.5cm (1–3in) long
Distribution	Worldwide
Habitat	Tree stumps on dry ground
Diet	Adults: nectar, fruit, and tree sap; larvae: decaying wood
Breeding	12–24 eggs

Praying Mantis

The fascinating praying mantises commonly kept as pets include the African (*Sphodromantis centralis*), Chinese (*Tenodera aridifolia sinensis*), dead leaf (*Deroplatys dessicata*), and Indian flower (*Creobroter meleagris*) mantises. The praying mantis gets its name from the manner in which resting mantises hold their forelegs, as if they are praying. Praying mantises are masters of camouflage, making use of protective coloration to blend in with foliage, and also mimic their surroundings with their form to appear like dead or living leaves, sticks, grass, flowers, or bark. Camouflage is not only a defense but also allows praying mantises to be excellent ambush predators. Praying mantises should usually be kept singly because they are cannibalistic. A terrarium planted with foliage and sticks for climbing, and furnished with a small water dish, should be provided.

Scientific name	*Sphodromantis* and other genera
Family	Mantidae
Size	1.2–15cm (0.5–6in) long
Distribution	Tropical and subtropical regions
Habitat	On vegetation
Diet	Insects such as crickets, moths, and flies
Breeding	Up to 400 eggs in a clutch

Ant Lion

Ant lion larvae make fascinating children's pets because of their ingenious trap-building activities. These larvae earn their name by their predatory behavior toward ants and other small insects. The ant lion larva creates a pit of 2.5–8 centimetres (1–3 inches) in diameter by digging backward in sand or sandy soil. In the process, spiral-shaped doodles are made, earning ant lions their nickname of "doodlebug." The ant lion then waits at the bottom of the pit for an insect to fall into it, and then pulls it under the sand. An antlion remains as a larva for up to three years, spending only a few weeks as an adult, in a form that resembles a dragonfly. Child-friendly kits complete with instructions, lidded dens, sand, and ant lions can be purchased online or from some pet stores.

Scientific name	*Myrmeleon crudelis* and other species
Family	Myrmeleontidae
Size	0.1–1.3cm (0.05–0.5in) long
Distribution	Mainly tropical and subtropical regions
Habitat	Mainly sandy soil
Diet	Larvae feed on ants, flightless fruit flies, and pinhead crickets
Breeding	Eggs laid in sand or soil

Giant Millipede

The giant millipedes commonly kept as pets include the black-and-red-striped Madagascan fire millipedes; giant black African millipedes, which reach lengths of 28 centimetres (11 inches); and Madagascan orange-legged millipedes, which, unlike these others, do not secrete a potentially dangerous stain on your fingers when they are handled. Giant millipedes are found mainly in the tropics. Like all millipedes, they have between 100 and 400 legs, with two legs per body segment. They spend most of their time in the soil, but will appear frequently, and some, such as the orange-legged, like to climb branches. The care and housing of millipedes depends on the species, but a captive environment should mimic their natural forest floor habitat as closely as possible. A large and humid terrarium with a soil substrate is recommended for most giant millipedes.

Scientific name	*Aphistogoniulus, Archispirostreptus,* and other genera
Family	Pachybolidae, Spirostreptidae, and other families
Size	Up to 30cm (12in) long
Distribution	Mainly in tropical regions
Habitat	In soil
Diet	Leaves, fruits, and vegetables, depending on species
Breeding	100 or more eggs laid in soil

Linnaeus' Leaf Insect

One of many species of leaf insects, or walking leaves, the Linnaeus' leaf bug has turned camouflage into an art form. To further confuse matters, this leaf bug rocks back and forth when it walks, to mimic a leaf being blown by the wind. When at rest, its legs and head lie flat against the abdomen to offer the most leaflike shape. If threatened, the bug will rub its wings together, making a rustling sound to scare away predators. Getting the housing just right is essential when caring for this relatively demanding species: the bug requires a plastic or glass container with a few air holes, kept at a year-round temperature of 20–25°C (68–77°F). Daily misting and placing peat on the bottom of the container will help maintain a suitable humidity.

Scientific name	*Phyllium siccifolium*
Family	Phylliidae
Size	6–7cm (2.4–2.75in) long
Distribution	Western Malaysia
Habitat	Arboreal
Diet	Bramble, wax myrtle, and salmonberry leaves
Breeding	Eggs take 5–6 months to hatch

Rhinoceros Beetle

The rhinoceros beetles, of which there are numerous genera and species, are in the scarab beetle family. They are among the largest beetles, and their common name refers to the horns of the males in most species, which are used in battles with other males. Rhinoceros beetles are also the world's strongest animals in relation to their size, with the ability to lift up to 850 times their weight. Popular pets are the Hercules beetle (*Dynastes hercules*) and Grant's rhinoceros beetle (*Dynastes granti*). For housing multiple beetles, a 110-litre (24-gallon) tank will be needed, with a secure, ventilated lid, as this species can fly. The feeding and care of each species differs, but the tank should usually be filled with 10 centimetres (4 inches) of soil, or soil that is mixed with mulch and leaves.

Scientific name	Dynastinae subfamily
Family	Scarabaeidae
Size	2–17cm (0.8–6.7in) long
Distribution	Worldwide
Habitat	Forest
Diet	Peeled juicy fruits, watered-down maple syrup, depending on species
Breeding	50–100 eggs

Dung Beetle

Dung beetles are found in numerous genera, but the beetle most popular among hobbyists, the Egyptian scarab (*Scarabeus sacer*), is found in the genus *Scarabeus*. This dung beetle, along with other North African species, was important to the ancient Egyptians, probably due to its feeding habits. Larvae feed on undigested plant material in dung, while adults get all the nourishment they need by extracting juice from the dung. The beetles search out dung using their powerful sense of smell, then roll it quickly away to avoid its being stolen by another beetle, before burying it as a food store or as a brooding ball. To the Egyptians, this rolling and burying symbolized the god Khepri, who rolled the sun across the sky before carrying it through to the other world.

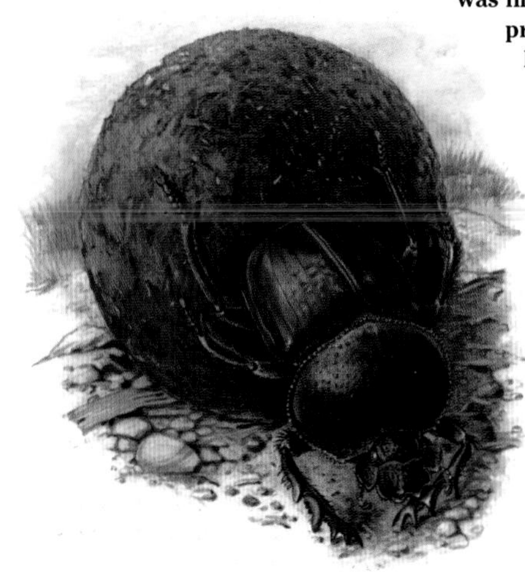

Scientific name	*Scarabeus* genus
Family	Scarabaeidae
Size	0.25–5cm (0.1–2in) long
Distribution	Worldwide
Habitat	Open countryside around herbivorous animals
Diet	Dung, preferably of herbivores
Breeding	Up to 3 eggs a day

Black Emperor Scorpion

Despite its fearsome name, this scorpion is one of the more gentle species and does well in captivity, making it a popular choice for beginners. Those who have been stung by the emperor scorpion compare it to nothing more than a bee sting. However, anyone—and children in particular—can be hypersensitive to the sting, which can have fatal results. In the wild, these scorpions live in groups within interconnecting tunnels, hunting on the rain-forest floor for insects and small vertebrates, which are seized with the scorpion's claws. Young scorpions, or scorplings, are born live and will ride around on their mother's back until their first molting. The emperor scorpion will need a humid 60-litre (13-gallon) tank with at least 8 centimetres (3 inches) of substrate and numerous rocks under which to hide.

Scientific name	*Pandinus imperator*
Family	Scorpionidae
Size	10–20cm (4–8in) long
Distribution	West Africa
Habitat	Damp floor of tropical rain forests
Diet	Anoles, cockroaches, crickets, and pink mice
Breeding	Up to 12 live scorplings

Mexican Red-Kneed Tarantula

This tarantula has a stocky, dark brown body and orange patches on its leg joints. The patella, or second leg joint, is entirely orange-red. Due to its docility, size, and attractive coloration, this tarantula is very popular in the pet trade. This high demand has led to the tarantula becoming classified as near threatened, so all pets purchased should be bred in captivity. Males may live for up to ten years in captivity, while females can survive for twenty-five years. A roomy terrarium mimicking the tarantula's woodland and savanna habitat should be provided. Since this tarantula is a burrower, it needs a deep sandy substrate. If irritated, the red-kneed tarantula will rub off the hairs on its abdomen, showering you with them. These can be extremely itchy and need to be rinsed off right away.

Scientific name	*Brachypelma smithi*
Family	Theraphosidae
Size	Leg span up to 17cm (6.7in)
Distribution	Western Pacific Sierra Madre Mountains, Mexico
Habitat	Scrub and forest
Diet	Moths, crickets, silkworm caterpillars, and an occasional pink mouse
Breeding	400–800 eggs laid in a burrow

King Baboon Tarantula

One of Africa's largest spiders, this tarantula is popular among hobbyists despite—or perhaps because of—its aggression. This pet is certainly not suitable for handling because it will rear up on its back legs if it perceives any threat and display its fangs, with which it will happily administer a nasty bite. This tarantula is rusty brown to rich brown and is covered with fine hair that makes it look velvety.

In the wild, the king baboon makes large underground burrows, where it can shelter until it is ready to ambush prey. In captivity, a king baboon should be housed singly in a tank of 60–120 litres (13–26 gallons) with 12–18 centimetres (5–7 inches) of bedding for burrowing. The tank will also need a locking screen top and a heat pad underneath.

Scientific name	*Citharischius crawshayi*
Family	Theraphosidae
Size	Leg span up to 23cm (9in)
Distribution	Kenya, Tanzania, and Uganda
Habitat	Dry scrubland
Diet	Mealworms, crickets, silkworm caterpillars, and an occasional pink mouse
Breeding	400–800 eggs laid in a burrow

Goliath Bird-Eating Tarantula

This aggressive tarantula, the largest spider in the world, is best kept by experienced hobbyists. Despite its name, the Goliath does not normally eat birds: its regular diet is insects and small vertebrates. When it does eat a bird, it will flip upside down in the bird's nest, so that the victim will land directly in its grasp. When threatened, the Goliath will stridulate, or rub its legs together, to produce a loud hissing sound. Like all tarantulas, it has fangs large enough to break the skin, but its bites are largely harmless to humans. It requires a large tank with enough substrate for burrowing. It is safest for the tank to be fairly shallow because this spi[illegible]ikes to cli[illegible]nd even [illegible]rt fall could kill it.

Scientific name	*Theraposa blondi*
Family	Theraphosidae
Size	Leg span up to 30cm (12in)
Distribution	Northeastern South America
Habitat	Marshy and swampy areas of rain forest
Diet	Insects and small animals such as pink mice and lizards
Breeding	100–400 eggs laid in a burrow

Glossary

Amphibian: An animal that can live part of its life on land and part in water, with a juvenile water-breathing form metamorphosing into an adult air-breathing form.

Arachnid: An invertebrate with a segmented body, an external skeleton, and four pairs of jointed legs, such as a spider or scorpion.

Bay: A coat color of horses that is dark reddish-brown.

Bicolor: A coat pattern with areas of one solid color in combination with white.

Blue: A coat color of blue-gray.

Brackish: Water that is less salty than normal seawater, as in an estuary.

Breed: A group of domestic animals within a species that have similar appearance, temperament, and skills.

Brindle: A coat pattern of subtle tiger stripes, often on a tawny or gray base.

Carapace: The hard upper shell of a tortoise, turtle, or terrapin.

Chameleon: A lizard with a grasping tail, long tongue, and the ability to change color.

CITES (Convention on International Trade in Endangered Species of Wild Fauna and Flora): An international agreement to ensure that trade in wild species doesn't threaten their survival.

Cobby: A short, stocky body shape.

Crustacean: An invertebrate with a hard shell and a segmented body.

Dorsal: Relating to the back or upper side.

Family: In biological classification, a family consists of a group of related genera (*see* genus).

Fetlock: Part of the back of a horse's leg, above the hoof, where a tuft of hair grows.

Freshwater: Water that is not salty, such as is found in lakes and rivers.

Genus (plural genera): In biological classification, a group of species having similar characteristics. In the scientific name for an animal, the genus comes first, followed by the species. For example, the domestic dog, *Canis familiaris*, is the species *familiaris* in the genus *Canis*.

Gill: Respiratory organ of fish through which they obtain oxygen from the water and get rid of carbon dioxide.

Habitat: The type of environment that an animal naturally inhabits, such as a coral reef or rain forest.

Invertebrate: An animal without a backbone.

Kennel club: A national organization with responsibility for the registration of pedigree dogs.

Larva: An immature form of an insect or amphibian.

Long-haired cat: Cats are primarily classified according to their coat length. The coat of a long-haired cat is profuse and medium to long.

Mammal: A vertebrate that bears live young, which is fed on its mother's milk.

Marine: Found in, or related to, the sea or salty water.

Marsupial: A mammal that is born incompletely developed and is usually suckled in a pouch on its mother's belly.

Omnivore: An animal that feeds on many kinds of food, usually on both plants and flesh.

Order: In biological classification, an order consists of a group of related families.

Oriental cat: A cat of fine-boned, lithe appearance, such as a Siamese.

Palomino: A coat color in horses, with a cream or golden body and a light-colored tail or mane.

Pedigree: Line of descent of a purebred animal, usually a dog, cat, or horse, showing the ancestry over a number of generations.

Plankton: Any plant or animal, often microscopic, that drifts or floats in water.

Pointed: A coat pattern in which the head, ears, legs, and tail are darker.

Pony: A horse of any small breed. Officially, a pony measures less than 14.2 hands (144 centimetres; 58 inches) at the withers.

Purebred: Refers to an animal that is the result of controlled pairings over a number of generations.

Reptile: An air-breathing vertebrate with skin covered in scales. Most reptiles are egg-layers.

Roan: A coat coloring of horses in which the main color is interspersed with hairs of another color, often bay or chestnut mixed with white or gray.

Scute: A bony external plate or scale.

Short-haired cat: Cats are primarily classified according to their coat length. The coat of a short-haired cat allows the underlying body shape to be seen.

Species: In biological classification, a species consists of similar individuals capable of interbreeding. A subspecies may show minor variations and inhabit a particular part of the species' geographical range.

Tabby: A coat marking on cats that can be striped, blotched, spotted, or ticked.

Tarantula: A large, hairy tropical spider of the family Theraposidae.

Temperate: Geographical regions with mild temperatures that lie between the tropical and polar zones.

Terrapin: A turtle that lives in brackish water.

Terrarium: A container with at least one transparent side in which small land animals are kept in conditions that mimic the ecosystem and temperature of their natural habitat.

Ticking: A coat pattern in which bands of color are seen on the hairs.

Tortoise: A land-dwelling turtle. In the United States, many terrestrial turtles are known as turtles or box turtles.

Tortoiseshell: A coat coloring, usually of cats, showing black as well as light and dark red areas.

Tropical: Refers to the warm regions lying between the tropics of Cancer and Capricorn, north and south of the equator.

Tubercle: A wartlike growth.

Turtle: A reptile with a body shielded by a hard shell.

Vertebrate: An animal with a backbone.

Index